Space: America's New Strategic Front Line

Nonproliferation Policy Education Center

Nonproliferation Policy Education Center
Arlington, VA 22216
www.npolicy.org

Printed in the United States of America.

ISBN: 978-1-7371113-3-7

Cover image: The third Space-Based Infrared Systems Geosynchronous Earth Orbit spacecraft launches aboard an Atlas V rocket (source: United Launch Alliance).

Nonproliferation Policy Education Center

The Nonproliferation Policy Education Center (NPEC), a 501(c)3 nonprofit organization, is a nonpartisan, educational organization founded in 1994 to promote a better understanding of strategic weapons proliferation issues. NPEC educates policymakers, journalists, and university professors about proliferation threats and possible new policies and measures to meet them.

For current publications of the Nonproliferation Policy Education Center, please visit www.npolicy.org.

Contents

Acknowledgments

This research was supported by generous grants from the Sarah Scaife Foundation and the Smith Richardson Foundation.

SPACE: AMERICA'S NEW STRATEGIC FRONT LINE

Henry Sokolski

America is steadily recognizing that space is a critical front line for strategic deterrence. The Pentagon has made space a new warfighting domain, created a U.S. Space Force, reestablished a U.S. Space Command, and supported allied efforts to create similar military space institutions. The question now is what such organizations should do.

One popular idea is that the United States and its space-faring partners should proliferate thousands of dual-use and dedicated military low-Earth orbit satellites that would be difficult for China or Russia to knockout. To reduce these constellations' cost and speed their development, the Pentagon would ask private space companies to assume much of this work. These satellite constellations would look down to earthly commercial and military operations. To help assure their continued operation in wars, the U.S. government would encourage commercial firms to reconstitute these constellations quickly with cheap, reusable launch systems. These efforts would afford a passive form of defense against anti-satellite attacks and obviate the need for destructive, active space defense systems.

Another view, compatible with this proliferated space architecture, is making space a sanctuary in which the United States and its allies do all they can to prevent it from being "militarized." Washington, it is argued, should encourage more international agreements with other space-faring nations — including China and Russia — to limit or ban anti-satellite operations; increase notice and data sharing on space launches, satellite locations, and operations; and establish rules regarding accidents and military liability. While attempting to secure such agreements, Washington and its partners should also build resilient satellite constellations that could passively survive possible attacks.

A final view that contrasts with these favors deterring space combat by establishing American military, civil, and commercial space superiority. This entails creating superior space launch capacity to launch payloads in all orbits; enabling satellites to transit to new orbits quickly; and being able to disable, dazzle and, if necessary, destroy hostile spacecraft. This hawkish view is generally paired with an expansive view of space commerce and exploration to assure both increased economic prosperity and to secure military "high" ground in cislunar space and beyond. It also is a view held by those wary of relying on space arms control.

Each of these views has their backers; none is entirely without merit. This prompts a series of questions. What is the optimal mix? Which one should be emphasized more than the other, at what time, and for how long? Which actions called for by each should be privatized and to what extent? Which should receive government support, to what extent, and in what form? Does it make sense to see space as a truly separate military domain like sea, land, and air? Or is it more appropriate to view space as an arena that lends support to warfighters conducting terrestrial combat?

Any effort to focus on the further development of U.S. and allied space capabilities, especially with an eye on deterring wars, must grapple with these issues. This volume is designed to help. It is broken into four sections.

The first focuses on how to think about the different schools of space strategy. Brent Ziarnick of the Air War University gives us an overview of how military thinkers have approached space deterrence and how it relates to combat on Earth. He offers a critique of those that attempt to apply nuclear deterrence theories to space. Peter Hays of George Washington University follows with a review of the most prominent views on deterring space combat. He makes the case for pursuing space superiority. This brings us to Pete Garretson's work. He details alternative budgets for a brown-water (downward-looking, Earth-servicing) space satellite force and a blue-water (expansive outwardly bound) space force system.

The volume's next section focuses on the emerging space activities of America's key Asian allies — South Korea and Japan — and its key adversary — China. Namrata Goswami details just how Beijing is challenging U.S. and allied command of space. Sam Wilson explains Japan's and South Korea's space plans.

This brings us to the volume's third section – a collection of studies briefed to the participants of three NPEC space wargames. The first of these studies is by Brian Chow and Brandon Kelley on emerging non-nuclear anti-satellite warfare threats — rendezvous satellite operations, ground-based lasers, and electronic warfare operations. This chapter details the threats and describes how best to cope with each militarily and diplomatically. The second chapter is by Greg Jones. It details the likely consequences of a nuclear explosion in space. This section's last study is by Laura Montgomery. It focuses on the ambiguities in the Outer Space Treaty, including the kind of orbiting weapons that are covered and what space weapons are not.

The volume's final section contains the NPEC space wargame after-action reports. The first is a wargame in which China blockades Taiwan and interferes with Japanese and South Korean satellites in an attempt to keep Japan and South Korea from helping Taiwan. The game ends with China attacking key U.S. military satellites. The second wargame has North Korea exploding a nuclear weapon in low-Earth orbit. One of the goals of this game is to understand the limits of international space law and the security risks of depending too heavily on low-Earth orbit satellite systems. The last wargame has Pakistani terrorists and India using dual-use Chinese and U.S. satellite services to strike military ground targets. This, in turn, catalyzes the United States and China to engage in space combat.

All of these games were designed to help determine if a "space Lusitania" is possible — a conflict in which the use of dual-use space services by lesser states could catalyze into a strategic conflict between major powers. Each validated that such escalation indeed was possible. This, in turn, suggests that space is, in fact, a separate military domain — a military theater, like those on Earth in which combat can escalate to total war. It also suggests that space is becoming a front line both of strategic combat and deterrence.

PART I

How Space-power Theory Affects One's Deterrence Outlook

Brent D. Ziarnick

Executive Summary

Even though there has never been a kinetic war in space, a "conventional wisdom" that emerged decades ago has now fossilized into dogma. The "Sanctuary school" of space-power believes that the space domain should be free of any weapons because space warfare is exceptionally dangerous for many reasons – including the danger from space debris and the potential for uncontrolled escalation into nuclear war on Earth. Almost all space defense policy, scholarship, and scholars uncritically follow this Sanctuary pattern of thought. However, few scholars have bothered to consider the assumptions that form the foundation of the Sanctuary school. Sanctuary assumptions are based on two questionable premises: that it is acceptable to transpose nuclear deterrence concepts directly and uncritically into the space domain and to assume that the technical limitations consistent with historical satellite systems are a permanent characteristic of the space domain itself. If the Sanctuary school's assumptions are correct, then modern Sanctuary adherents' conventional recommendations may be useful. However, if Sanctuary assumptions are incorrect, then new analyses founded on better assumptions are necessary to understand the strategic reality of the space domain.

This paper explores space deterrence from a different departure point than the Sanctuary school – space-power theory. The paper explores Sanctuary school concepts borrowed from nuclear deterrence theory such as escalation and first-strike stability and assesses their utility in the space domain, highlighting their weaknesses and the consequences of Sanctuary thought on space deterrence, culminating in the new emphasis on so-called "Cross Domain Deterrence."

Next, the paper offers a clean-slate discussion of space deterrence departing from space-power theory, especially its insight into space control as a critical component of space deterrence (as control is a critical component of deterrence in all other domains of warfare), the nature of space warfare as a strictly limited form of warfare contra the Sanctuary school, and space-power theory's assertion that defense is still the stronger form of war in the space domain. From these foundations, a new approach to space deterrence – dynamic space deterrence, is offered that acknowledges space-power theory's conclusions on the nature of space-power. The paper then offers two historical analogies that help describe dynamic space deterrence. The first is a discussion of how Britain's unparalleled maritime industrial/technical/military supremacy acted as a deterrent force during the "Pax Britannica" of the 18th century and how space forces can duplicate the Royal Navy's effects. Second, World War One airpower is used to differentiate the Sanctuary school's "Balloon Mentality" of space warfare versus space-power theory's "Airplane Mentality" stressing movement and maneuver in force design for effective space warfighting and durable deterrence. Finally, the paper expands on the technologies and mindsets needed for dynamic space deterrence and offers a vision of limited space warfare.

Decades of scholarship on space-power theory suggest new methods of examining space issues that are often ignored by conventional space policy scholars firmly in the Sanctuary tradition. This paper offers that space-power theory indicates that space deterrence – the prevention of kinetic war in outer space – may be best supported by pursuing the exact opposite of Sanctuary recommendations. Only the development and deployment of advanced space warfighting capabilities can support a credible space deterrence posture. Combined with the possibilities of advanced technologies that seem likely appear in the near term, this study suggests that the nature of space-power may be significantly different than what Sanctuary theorists imply and offers a number of unique insights to advance the space deterrence discussion beyond the Sanctuary school.

Introduction

The relationship between space activity and deterrence has been debated since the early Cold War. In the 1950s and 1960s the debate centered on the strategic nuclear-armed intercontinental ballistic missile (ICBM). In the 1980s, the focus shifted to the role of the strategic satellite for early warning and nuclear command and control. In the late 1908s there was a brief flirtation with space-based missile defense during the Strategic Defense Initiative (SDI) program. However, with the end of the Cold War and the American victory in the 1991 Gulf War, space came to be seen as an enabler for warfighting rather than as a strategic asset best suited to nuclear deterrence. Since then, space forces have been developed and prized for their ability to provide space support to terrestrial warfighting at the operational and, increasingly, tactical levels – in addition to being vital infrastructure for the globalized economy. Because space forces are now seen more as an active participant in tactical warfighting and economic health, the relationship between space and deterrence has shifted from its earlier Cold War assumptions.

For the last 30 years, the space and deterrence discussion has largely been how to deter war *in* space (space deterrence) rather than how space can contribute to nuclear deterrence. However, this critical question is hampered by its nuclear deterrence past in two major ways. Firstly, the ideas developed around nuclear deterrence are often applied to space without critical reflection on whether nuclear deterrence concepts remain relevant in discussing how *conventional* deterrence works, even in a very *unconventional* domain like space. Secondly, because the accepted space deterrence literature has been written primarily by those who study nuclear deterrence rather than space strategy and operations, the uniqueness of the space domain is either overexaggerated to serve political agendas (such as the "Kessler Syndrome" and the space debris threat) or ignored altogether. These nuclear assumptions have coalesced into the "Sanctuary school" of space, that argues that space weapons (like nuclear weapons) are uniquely dangerous and lead to uncontrollable escalation into nuclear war. Because of these cataclysmic assumptions, little measured and politically neutral exploration of the utility of space weaponization at the operational and strategic levels of war, or space deterrence, have been attempted, leaving a dearth of options regarding space strategy.

Space-power theory might prove to be a useful and needed corrective solve the problem of space deterrence. By focusing on the strategic attributes of the space domain itself, new insights into deterring war in space and, by extension, keeping the peace and minimizing terrestrial state-on-state violence, can help leaders guide the international community into ever-increasing peace and prosperity on Earth through human expansion into space. But before space-power theory can be applied to deterrence, the pitfalls of nuclear thinking applied to space must be exposed.

The Pitfalls of "Traditional" Nuclear Deterrence

Most academic studies on deterrence in the Cold War focused on nuclear weapons and deterring a cataclysmic nuclear war and, to a lesser extent, large-scale conventional war in Europe through nuclear retaliation threats. Given the horribly destructive nature of nuclear weapons, Cold War deterrence placed a premium on deterring *any* type of nuclear warfare among *any* belligerents, worrying that any use of a nuclear weapon in war would immediately escalate into a world-ending confrontation. The concepts of "stability" and "first-strike stability" to resist escalation were developed to come to terms with the threat of nuclear forces capable of destroying human civilization that made any conceivable violent use of nuclear weapons supremely risky to the point of insanity.

While these assumptions may have been valid for nuclear weapons, it is far less clear if such concepts are useful for weapons that, if used by themselves, post no immediate threat to end all life on Earth. Regardless, in the space deterrence literature, many nuclear deterrence concepts have been applied by nuclear deterrence academics without seriously reflection on the validity of their assumptions in the very different domain of space. The fundamental problem in applying nuclear deterrence concepts to the space domain is the assumption of inevitable escalation to nuclear warfare after the use of space weapons.

Escalation and the Sanctuary School

The "Sanctuary school" of space emerged in the 1970s from the idealism of the Outer Space Treaty of 1967, that promoted the "peaceful use of outer space," and the fear of nuclear war. The Sanctuary school's logic was rooted in the logic of nuclear deterrence. The goal of the Sanctuary school was to keep space free of weapons and free of wars. The Sanctuary school believed that the primary value of space forces – which they limit essentially to military support satellites - is their ability to reduce the probability of global nuclear war through their observation, communication, and especially early warning capabilities. Because the school developed well before the 1991 Gulf War and the emergence of space support forces as terrestrial warfighting force multipliers, Sanctuary adherents saw no value in war in space other than as an adjunct to nuclear war, especially as a prelude to a nuclear first strike. "At all levels of conflict short of global nuclear war, space forces retain their value only if the environment remains a sanctuary," advocates argued.[1] If one nation loses the ability to see into the other side's territory through the loss of missile warning or other spy satellites, the blinded faction may see no choice other than to escalate to general war and, if it possesses them, use of nuclear weapons. If one side destroys the other's missile warning and/or nuclear command and control satellites, the defending side may see the space attack as a prelude to a nuclear strike and retaliate immediately, for there are very few reasons to destroy these types of satellites in a limited war scenario. Destroying the ability of the adversary to know what is happening is begging for miscalculation and, in turn, paranoid escalation.[2]

Sanctuary school advocates retained the link between the space domain and nuclear war primarily for its utility to argue against space weapons, and not without some cause. Sanctuary thinkers argued that "the aggressor would have to worry about the effects of inadvertently destroying the other nation's assets that support the deterrent strategy" in a space strike that was anything less than absolutely surgical. They argued that "a show of force intended to

---

1. David Lupton, *On Space Warfare: A Space Power Doctrine* (Maxwell AFB, Alabama: Air University Press, 1988), 34-5.
2. David Lupton, *On Space Warfare*, 34-5.

destroy a satellite broadcasting propaganda might cause a global nuclear war if the attack inadvertently destroyed the opponent's ballistic missile attack warning system."[3]

The Sanctuary school proper has been generally minimized as obsolete by academics in the modern era. Modern deterrence theorists have concluded that it "is becoming increasingly possible to strike [space systems] that support conventional military capabilities without threatening the opponent's nuclear C2 infrastructure, ability to detect nuclear missile launches, or commercial or other independent satellite capabilities." There is wide agreement that modern space systems are no longer "tightly coupled to strategic nuclear stability" as they were in the Cold War.[4] However, the Sanctuary school successfully linked nuclear conflict to space warfare in the popular deterrence imagination, leading to other theorists retaining the same questionable space deterrence concepts borrowed from nuclear deterrence.

First Strike Stability

Even if the Sanctuary school itself has been widely dismissed, their linkage between nuclear and space deterrence persists – and remain an open avenue for other shaky ideas to creep into the discussion. Another such concept is that of "first strike stability." A critical factor in nuclear deterrence was believed to be the ability of a nation to successfully retaliate with sufficient nuclear force even if it suffered a catastrophic first-strike nuclear attack aimed at destroying their nuclear deterrence forces. If the defender could successfully

---

3. David Lupton, *On Space Warfare*, 56.
4. Benjamin W. Bahney, et al. " anti-satellite Weapons and the Growing Instability of Deterrence," in Jon R. Lindsay and Erik Gartzke, *Cross-Domain Deterrence: Strategy in an Era of Complexity* (New York, New York: Oxford University Press, 2019), pg 135-6.

survive the best first-strike an aggressor could reasonably mount sufficient to retaliate with unacceptable damage to the aggressor, the defender would successfully deter any adversary from a surprise nuclear attack altogether. Since first strike stability is such an important nuclear deterrence concept, it was perhaps inevitable that the concept would be applied to space deterrence.

RAND researcher Forrest Morgan developed perhaps the most advanced case for applying "first strike stability" in space. Morgan admitted that the immediacy and intensity of escalation in space war was not anywhere nearly as consequential as in nuclear war, but he insisted "several other strong parallels exist between first-strike stability in space and in the nuclear realm." First, Morgan argued that space support's force multiplying ability to conventional terrestrial forces are threatening to enemies and, therefore, space assets are tempting targets. Additionally, satellites are difficult to defend. "As a result," Morgan argues, "space, like the nuclear realm, is an offense-dominant environment with substantial incentives for striking first should war appear probable." Next, "deterrence failures in space, though not as immediately catastrophic as nuclear deterrence failures, could, nonetheless, be very costly given the resources invested in orbital infrastructure and the many security and economic functions that benefit from space support." Also, Morgan claims that, "like nuclear deterrence failures, the costs of warfare in space would likely be shared by third parties due to global economic interdependence and multinational ownership of many space systems—all the more so if kinetic attacks on satellites litter important orbits with debris." Finally, Morgan says, "there is a parallel between nuclear and space deterrence in that significant thresholds are perceived in both realms, the crossing of which could lead to reprisals, follow-on attacks, and rapid escalation."[5]

---

5. Forrest Morgan, *Deterrence and First-Strike Stability in Space: A Preliminary Assessment* (Santa Monica, California: RAND, 2010) , 2.

Morgan's conclusion does not follow from his analysis. While Morgan is right that current satellite systems provide critical support to terrestrial forces and satellites are currently difficult to defend, he concludes that space is therefore an offense-dominant domain in all important facets equivalent to the nuclear arena. Stephen van Evera defined "offense dominant" as "conquest is fairly easy" or "offense is easier than usual" while "defense dominant" meant "conquest is very difficult."[6] Morgan simply assumed that the space domain was offense-dominant because satellites have never been built with any defenses - not on the nature of the space domain itself - so his conclusion that space is an offense-dominant domain where striking first is highly prized is, at best, uncertain. However, Morgan went even further into wild speculation by equating a space war's real potential to disrupt national economies with cascading international effects and possible environmental harm of space debris with the potentially world-ending consequences of a Cold War-style general nuclear war. Here, Morgan slyly attempts to convince the reader that similarities in *kind* (both nuclear and space war effects may have international and environmental consequences) are enough to justify using nuclear assumptions in analyzing space war even though they are outrageously different in *degree* (in no way can space-focused war of any kind threaten immediate civilization destroying damage or the end of all life on Earth).

Assuming that a nuclear style first-strike stability is necessary for space deterrence is an error because it drives a space deterrent posture logic that holds stability as the greatest aspiration for defense space activity. Furthermore, retaining stability champions "static deterrence" through assured destruction as the answer to avoiding war in space. Static deterrence requires an overwhelming counterattack of such intensity and damage in scope for an attempt at space aggression sufficient to make any aggressor considering a

---

6. Steven van Evera, "Offense, Defense, and the Causes of War," *International Security*, Vol 22, No 4, (MIT Press, Spring 1998), pg 6.

space attack to immediately reject for fear of unacceptable consequence. In this construct, the key to deterrence is *stability*. Both sides must be comfortable in its first strike stability to maintain a Cold War nuclear-like "balance of terror" to prevent space warfare. In this environment, dynamic space activity – rapid advances in space operations in any arena - is threatening to stability and any aggression in space must be met with immediate escalation in order to keep the peace, even to the point of using nuclear weapons. The static deterrence posture of immediate escalation is antithetical the true strategic nature of the space domain, discussed below, and exceptionally dangerous to upholding international stability, forming the essential paradox of the Sanctuary school.

The Static Deterrence Cross Domain Response: A Time and Place of Our Choosing

For the Sanctuary theorist, space deterrence requires a static deterrence posture that enforces stability through responding to any act of space aggression through immediate retaliation with overwhelming, unacceptable force completely disproportionate to any space attack. This erroneous use of the nuclear first strike stability concept has severely distorted thinking about the space domain. The Sanctuary school's fundamental misunderstanding of the nature of space-to-space weapons may have been academically discredited, but it still holds great sway across the political and defense establishments in the United States. No space-to-space weapons have been built or deployed as yet by America's armed forces. Since space weapons are politically verboten and acknowledged attacks on space systems require an overwhelming response resulting in unacceptable damage to an attacker for static deterrence to work, the question of where this overwhelming force is supposed to come from has been questioned.

Thus enters the concept of Cross Domain Deterrence. Cross Domain Deterrence (CDD) is "the use of threats of one type, or some combination of different types, to dissuade a target from taking actions of another type to attempt to change the status quo. More simply, CDD is the use of unlike technological means for the political ends of deterrence."[7] Although CDD has been only recently intellectually matured by academia, the (unnamed) concept was expressed in official government guidance much earlier.

The 2011 National Security Space Strategy (NSSS) declared that the United States sought "to address the *contested* environment with a multilayered deterrence approach... We will retain the right and capabilities to respond in self-defense, should deterrence fail."[8] Without space weapons, however, the question remained how the US planned to respond in self-defense. The document continued, "Partnerships with other nations, commercial firms, and international organizations, as well as alternative U.S. Government approaches such as cross-domain solutions, hosted payloads, responsive options, and other innovative solutions, can deliver capability, should our space systems be attacked… Finally, the United States will retain the right and capabilities to respond in self-defense, should deterrence fail. We will use force in a manner that is consistent with longstanding principles of international law, treaties to which the United States is a party, and the inherent right of self-defense."[9]

While the NSSS was vague, an article published concurrently with the document by James Finch and Shawn Steene, the director and deputy director of Space Policy and Strategy Development in the Office of the U.S. Undersecretary of Defense for Policy, argued that

---

7. Jon R. Lindsay and Erik Gartzke, *Cross-Domain Deterrence: Strategy in an Era of Complexity* (New York, New York: Oxford University Press, 2019), 4.
8. United States Government, *National Security Space Strategy Unclassified Summary*, January 2011, 13.
9. *National Security Space Strategy,* 10.

"the credible ability to carry our retaliation [for attacks against space systems]... includes all elements of national power in any domain. It need not, and should not, be limited to military actions in the space domain."[10] What was vague in 2011 became explicit policy in the 2020 National Space Policy. It stated, "Any purposeful interference with or an attack upon the space systems of the United States or its allies that directly affects national rights will be met with a deliberate response at a time, place, manner, and domain of our choosing."[11] It expanded on this declaration a few pages later with the "United States will develop strategies, capabilities, and options to respond to any purposeful interference with or attack on the space systems of the United States or its allies that directly affects national rights, especially those necessary for the operation of the Nation's critical infrastructure. Such strategies, capabilities, and options will allow for a deliberate response at a time, place, manner, and domain of its choosing."[12] CDD is declaratory US national policy for space deterrence.

CDD is a logical solution to the space deterrence problem interpreted through the lens of nuclear deterrence theory. CDD advocates parrot Morgan's argument that "Irresponsible acts against space systems [which a space attack would be since it has no logical purpose, per the Sanctuary school] could have implications beyond the space domain, disrupting worldwide services upon which the civil and commercial sectors depend."[13] Thus, a space attack must be deterred with escalatory overwhelming reprisal to inflict unacceptable cost to the attacker. CDD allows that force to come from any source.

---

10. Bahney, et al. " anti-satellite Weapons," 138.
11. United States Government, *National Space Policy of the United States of America,* 9 December 2020, 4. Emphasis original.
12. *National Space Policy,* 9.
13. Bahney, et al. " anti-satellite Weapons," 122-3.

Even a limited consideration of CDD reveals worrisome incongruities. Static, CDD-based space deterrence seeks to avoid the risk of uncontrollable escalation to nuclear war that they believe is inherent in space war by threatening the same devastation for any aggression in space. Thus, static space deterrence will itself spark the massive escalation it is meant to prevent if it fails to deter any and every potential space attack. Worse, however, is that most space-on-space attacks would not directly threaten any human life at all (or at most a handful of astronauts), but the reprisal from CDD space deterrence would likely cause untold casualties and massive human suffering. The phrase "satellites don't have mothers" was a popular response in military space circles after reading that the US planned to destroy ground targets with anticipated collateral deaths in reprisal for an attack on a satellite. The moral questionability of this policy was so profound that many thought even a limited terrestrial reprisal threat lacked credibility, much less the overwhelming retaliation envisioned by CDD advocates.

It is notable, but perhaps not surprising, that CDD theorists who were supposed to be experts at nuclear deterrence theory were unaware that a great nuclear theorist wrote against deterrence across domains almost seventy years ago. In his 1966 classic *Arms and Influence*, Thomas Schelling considered the limited aspects of the Gulf of Tonkin incident and the U.S.' response. Schelling explained how the Gulf of Tonkin did not immediately escalate through his concept of the "idiom of reprisal." "It seems that governments do feel obliged to make a pattern of their actions," Schelling argued, "to communicate with deeds as well as with words." He continued "There is something here in the psychology of communication – in people's sense of proportion, of justice, or appropriateness… that goes beyond the abstract military relation between enemies, beyond the economics of cost and damage, beyond the words that are used to rationalize a set of actions… There is an *idiom* in this interaction, a tendency to keep things in the same currency, to respond in the same language, to make the punishment fit the character of the crime, to impose a coherent pattern on relations."[14]

It is absolutely clear that CDD's demand that the initiator of any space attack, however limited and free of casualties the strike was, suffer immediate reprisal with overwhelming air, land, or sea force to inflict unacceptable damage to the attacker – even to massive loss of life – is a gross violation of Schelling's idiom of reprisal. It is morally shocking to kill large amounts of people in their own territory for a limited war attack on an inanimate object, even if in space and even if it could cause moderate economic discomfort to the defending nation.

Perhaps more consequentially, however, is that modern CDD theorists see that new technologies are making the escalatory and violent reprisal demanded by nuclear-style deterrence illogical. One analyst wrote, "technological developments – particularly the development of non- or minimal debris-producing anti-satellite (ASAT) options, as well as reversible ASAT operations like jamming – will also strengthen the logic of ASAT strikes against space-enabled opponents… Put differently, future ASAT threats are likely to be perceived as consistent with limited war constructs rather than as precursors to nuclear escalation."[15] As will be discussed below, even the CDD theorists could not shake the feeling that space war was not really escalatory by nature like nuclear combat was. Even they felt that space war was rather an expression of limited war. But they could not explain why, because they knew nothing about actual space strategy or the nature of the space domain. It takes a space-power theorist to describe the true nature of space conflict and combat, and what the true requirements of space deterrence are.

---

14. Schelling, *Arms and Influence*, 146-7.
15. Bahney, et al. " anti-satellite Weapons," 136.

The Failure of Static Deterrence

The ultimate failure of applying nuclear deterrence concepts to the space domain is that its concept of CDD space deterrence attempts to deter *any* war in space with nuclear offense-dominant assumptions that require the same catastrophic escalation that so frightened Sanctuary theorists in the first place. The Sanctuary school thus created the very existential danger that it accuses space warfare of threatening. The Sanctuary school offers only a "self-fulfilling prophecy" based on its own fallacious assumptions. However, the space domain is nothing like the nuclear arena and need not threaten uncontrollable escalation at all. Space is, in fact, a *defense*-dominant domain characterized by energetic dynamism where stability can never be achieved nor should be sought by rational strategy. Further, space warfare is an inherently *limited* form of war that, by nature, is non-escalatory because it is not in-itself directly or immediately lethal to a state or its citizenry.[16]

Because nuclear deterrence advocates have mischaracterized the space domain's strategic character so severely, their solutions to space deterrence are equally misguided and, potentially, extremely dangerous. By simply maintaining the erroneous link between space and nuclear warfare and suffering under the delusion that the space domain is an offense-dominant domain because they mistake the inadequacy of current satellite defensive technology for the characteristics of the domain itself, they risk a self-fulfilling prophesy. Suffering under their flawed logic, national leaders under space attack may feel obligated to retaliate using needlessly escalatory methods, including nuclear weapons because they may *feel* they are being threatened more than they actually are. Only by

---

16. See John Klein, *Space Warfare: Strategy, Principles and Policy* (London, United Kingdom: Routledge, 2006), 102 for his discussion on limited aims.

understanding the nature of the space domain through space-power theory can this danger be removed and true understanding achieved.

Space Deterrence and the Nature of Space Warfare

Space deterrence is intimately connected with space warfare. Violent force is critical to provide the capability that must be matched with credibility and communication necessary for effective deterrence in any form. While Schelling's idiom of reprisal suggests that constructive credibility and communication requires space weapons of some sort, space-power theory is necessary to determine what types of space weapons are required and how they should most effectively be used. Fortunately, space-power theory has richly explored the nature of violence in space-power.

The most important space-power theorist to explore the nature of space warfare is John Klein. His 2006 book *Space Warfare: Strategy, Principles, and Policy* followed much of previous space-power theory in accepting that space-power shares many common attributes to seapower, and Klein adapted British maritime theorist Julian Corbett to the space domain. Taking much of Corbett's strategic writing at face value, Klein concluded that the "inherent value of space is the utility and access it provides, and this utility and access are enabled through celestial lines of communication… those lines of communications in and through space used for the movement of trade, materiel, supplies, personnel, spacecraft, electromagnetic transmissions, and some military effects."[17] Klein's "celestial lines of communication" (CLOCs) concept was lifted from Corbett's sea lines of communication because Klein correctly determined that the strategic concept carried over extremely well.

---

17. John Klein, *Space Warfare: Strategy, Principles and Policy* (London, United Kingdom: Routledge, 2006), 51.

But the concepts were not completely alike. With the advent of modern communications through the electromagnetic spectrum, Klein adjusted Corbett's concept slightly and explained that there were both "physical" and "non-physical" CLOCs. Physical CLOCs were spacecraft and orbital trajectories and had direct analogies to the maritime domain. They were also tactile and easily understood. However, non-physical CLOCs were something fundamentally different than anything seen in classic seapower. "A key difference [between space and maritime LOCs] is that many of the things being communicated in space are not directly tangible in a conventional manner," Klein argued.[18] This key difference is from nonphysical communications that "comprise the transmission of data, information, and some military effects along celestial lines of communication."[19] The nonphysical CLOCs are the most prevalent today since much of space-power's utility is through the collection and distribution of data and communications via invisible electromagnetic signals.

Klein next described CLOCs as the central object of space warfare. Again, borrowing from Corbett, Klein argued that the "primary objective of space warfare is to protect and defend one's own lines of communications, while limiting the enemy's ability to use his."[20] The CLOC is the central currency of space-power and that which space warfare is fought to control. Space control, then, is control of the physical and nonphysical CLOCs important to the two belligerents. Klein's definition of space warfare is comprehensive of both today's space control activities (whether nonphysical CLOC attacks such as satellite downlink jamming or physical such as direct ascent kinetic missiles). Klein expands his theory to discuss cruiser warfare, blocking activities, and other advanced military space options based on control of CLOCs. There are many insights found

---

18. Klein, *Space Warfare*, 52.
19. Klein, *Space Warfare*, 53.
20. Klein, *Space Warfare*, 51.

in Klein's *Space Warfare* and later writings, but the CLOC is without question his most profound contribution to space-power theory.

An example of the utility and correctness of Klein's work is that current USSF space control missions fit easily in the CLOC space warfare model. Ground-to-space and space-to-space jammers, such as the USSF Counter-Communications System (CCS) disrupt and deny the non-physical CLOCs that allow information to flow between the spacecraft and terrestrial nodes such as users and control elements. Physical destructive weapons such as direct-ascent anti-satellite weapons are intended to destroy the space vehicle and eliminating access to the CLOC (indeed eliminating the CLOC itself) the satellite provides, an action against physical access and hence an attack on the physical CLOC.

Klein's insights also connect space to space warfare with limited war. Klein argued, "limited aims include only those attempting to acquire certain lines of communications in space or assuring one's access within a region of space, but *they do not include the overthrow of a sovereign government.*"[21] Since space warfare proper is the attack and defense of CLOCs, space warfare in isolation can only be fought to achieve limited aims. Terrestrial strike from space is a *use* of space-power, not an act of space warfare, and can be used (in conjunction with other means) as a method of unlimited war, but not space warfare itself.

---

21. Klein, *Space Warfare*, 102. Emphasis added.

Defense Is the Stronger Form of Space War

Carl von Clausewitz said that "*the defensive form of warfare is intrinsically stronger than the offensive.*"[22] Klein argued that despite "space warfare being different from the land warfare theory of which Clausewitz writes, the defense is the stronger form of war… *always*.[23] Klein offers that the defense stronger because the defender is already established in a position and "awaits the blow." These are advantages in the space domain as well as on land or sea.[24] Also, by advancing, attacking space systems gives away its position and disposition, inviting counterattack. Further, the attack may also provide essential clarity to the strategic intentions of the attacker, allowing the defender to respond appropriately at the attack in all other domains and dimensions in addition to the space attack.[25] Defense also gains strength from the fact that the defender always has the inherent right of self-defense, increasing the moral authority of the attacked as well as the right to reprisal, which can yield tremendous benefits in the larger conflict or diplomatic struggle in the attack's aftermath.[26]

Klein blames the common but erroneous assumption that space is an offense-dominant domain on space technology and the lack of defensive concern that designers have for space systems, not the domain itself. Defense is indeed the stronger form of space war, "but to successfully assume a defensive posture, an advantageous position must be held." Klein argues, "Placing space-based defensive systems in an unprotected, highly vulnerable location – as in a near-earth orbit – without any kind of relative advantage is not

---

22. Carl von Clausewitz, Michael Howard and Peter Paret (trans. and ed.), *On War* (Princeton, New Jersey, Princeton University Press, 1976), pg 358. Emphasis original.
23. Klein, *Space Warfare*, 75. Emphasis original.
24. Klein, *Space Warfare*, pg 75.
25. Klein, *Space Warfare*, pg 75.
26. Klein, *Space Warfare*, pg 76.

the strategy of the defense but is the *strategy of the ludicrous*."[27] Space systems have simply been built too weak and defenseless due to an overemphasis on cost over defensibility. Klein dismisses the offense-dominant argument by asserting that even though "space-based systems seem highly vulnerable to offensive weapons, this is a transitory state of affairs within the latest technology cycle."[28] Sooner or later, defensive technology will be developed to overcome the offensive advantages of space weapons. The race between technologies of offense and defense has always been, and ever shall be.

Another critical defensive strength of the space domain not mentioned by Klein but absolutely relevant is the sheer altitude of spacecraft with respect to Earth weapons, including the "high ground" and "gravity well" effects. From extremely high altitudes, space systems can "see" earth objects in wider swaths than earth systems can see space objects. Furthermore, the orbiting space object will likely be able to see physical weapons launched from Earth (such as missiles) extremely easy if sufficiently equipped. These height advantages are significant so long as space systems are designed to take advantage of them. Likewise, objects orbiting in space, through the potential energy granted by extreme altitudes, are much more energetic than objects that must "rise out" of the Earth's gravity well to reach the target. The gravity well may not significantly impact directed energy weapons (though other defensive methods are available), the energy requirements of an ascending weapon will accrue movement and maneuver advantages to the defender already in orbit.

The common mistake of believing space system fragility is a characteristic of the space domain rather than a sign of technological immaturity also leads to the overemphasis of space debris as a

27. Klein, *Space Warfare*, pg 75. Emphasis original.
28. Klein, *Space Warfare*, pg 78.

critical concern. Space debris is not a characteristic of space, nor a gun pointed at the head of the universe. Rather, the problem of space debris a function of human commitment, of choice. Humans have technology capable of cleaning up space, it just has not yet been built because of "static" space concerns born of the erroneous assumptions of nuclear deterrence discussed above. Fighting in space not unthinkable due to considerations of debris. Space debris, unlike many of the terrestrial domains, does not naturally clean itself through natural processes, though atmospheric drag does generally clear low Earth orbits of debris every eleven-year solar cycle. However, active debris removal systems using high energy lasers or physical methods can solve much of the debris problem, alongside better debris tracking and heavier spacecraft with additional shielding.

The space debris problem is, however, pushed by Sanctuary advocates who use this exaggerated threat to sell arms control policies to politicians and defense professionals. This misguided advocacy obscures the fact that space warfare is very *limited* in its escalation potential if it is focused on CLOCs. Thus, economics-based space deterrence strategies based on the threat of denial or punishment via attacks on CLOCs that are focused on limited means, ways, and ends do not require threats of vital interior, population, or civilizational damage. Space deterrence does not need to rely on uncontrollable escalation to be effective.

Space-power, like seapower, is dynamic and not suited to theories of nuclear deterrence that relies on ever-alert but static forces, including nuclear forces, on threats of uncontrollable escalation, or are prone to believing actual space warfare is "unthinkable." Rather, space warfare is a type of limited warfare that, if limited to the space domain, is likely to be relatively bloodless, especially compared to land warfare. Space warfare is also extremely dynamic, just like the space domain itself, where rapid change and innovation will check any attempt to keep the domain static and unchanging. Space warfare, especially its most limited forms such as jamming and minor physical attacks on solitary satellites or systems, will also probably occur increasingly often. Therefore, what is needed is a dynamic mode of space deterrence based on space-power theory. Space-power theory suggests domain control – the ability to operate with maximum freedom against the resistance of both the harsh domain itself as well as potential adversary action – is the key to dynamic space deterrence. A dynamic strategy that relies on domain mastery through space control to present the credibility, capability and communication for effective space deterrence requires space forces that are quantitatively and qualitatively superior at space warfare sufficient to deter any would-be aggressors. Such a force would be based around conventional space-to-space warfighting capability to ensure that the space domain's nature as a limited warfare tool concerned with attack and defense of CLOCs, and not terrestrial strike (which would be highly escalatory and eliminate space warfare's limited nature), is respected in order to limit escalation of violence on Earth. These systems would be based almost entirely in space to limit the potential of unstoppable escalation endemic to cross-domain strategies. This dynamic deterrence structure would be based on conventional forces' violent capabilities against enemy space forces and economic loss through CLOC harassment, a far cry from nuclear deterrence strategies. Fortunately, history offers a solid analogy for emulation.

Pax Britannica and the Royal Navy

In the mid-nineteenth century, Great Britain was the world's predominant sea power. It often used its Royal Navy to achieve its diplomatic and commercial goals. Its merchant marine transported British production across the globe and received staples such as tea, food, and raw materials to feed is growing population and ravenous industrial machine. The Royal Navy secured the critical sea lines of communication as well as used its power to threaten naval force to deter potential adversaries and actually used its powers against others. The technologically advanced Royal Navy both fought some conflicts and deterred others, but overall kept a relatively calm global peace. [29] Historian Rebecca Matzke argued that Britain's full-spectrum dominance of all aspects of seapower, their "commercial, industrial, financial, and imperial might, all protected by the superior Royal Navy, were so imposing that they deterred war, either by compelling other nations and peoples to defer to Britain's will, or at least by inviting them to bask in the safety afforded by a navy that upheld the status quo."[30]

Britain's unique role here is incredibly instructive for the United States. A multipolar world will likely stabilize into a world of regional hegemonies that are reasonably comfortable in their territories. Relative power gains in such a world will likely come through space control activities as great powers gain wealth and technology from strategic expansion into space. Focusing expansion on a limited war playing field such as outer space can capture the advantages that Great Britain had at sea: the United States can be an expansionist power by also maintaining the terrestrial status quo, provided it did not use its power to project violent force from space. Space powers must abandon terrestrial ambitions and focus on expanding their technological and economic dominance of space for commercial, political, and military gain.

An important lesson of this era is that "British officials thoughtfully considered new technology and adopted it when its abilities were proven, Matzke explains, but for "reasons of economy, though, they left much of the initial research and development to Britain's

29. Rebecca Berens Matzke, *Deterrence through Strength: British Naval Power and Foreign Policy under Pax Britannica* (Lincoln, Nebraska: University of Nebraska Press, 2011), pg. vii.
30. Matzke, *Deterrence through Strength*, 1.

thriving private sector. The many changes embraced by the Admiralty preserved the Royal Navy's position as the strongest and most effective navy in the world."[31] This element of the Royal Navy's supremacy is essential to comprehend for the space domain. Today, the commercial sector, especially its most entrepreneurial, is the engine of space innovation. If the U.S. Space Force lacks the budget for rapid space innovation, the example of the Royal Navy's use of its robust commercial sector for technology development is an important lesson.

Another lesson from the Pax Britannica is what class of technology might be most critical for space control – movement and maneuver. "Perhaps the most important new technology for increasing the capabilities and reach of the British Navy," according to Matzke, "was steam power. Between 1815 and 1850 Britain pioneered the development of steam warships and the tactics for using them."[32]

The HEICS *Nemesis* is an appropriate symbol for the Royal Navy's full spectrum dominance of seapower during the Pax Britannica. Built for the East India Company, not the Royal Navy, *Nemesis* was the first iron-hulled steam powered warship in the world. The East India Company had used steamships for war since the 1820's due to their shallower draft and exact navigation which allowed for riverine operations unsuited to sailing warships. *Nemesis* entered service during the First Opium War against China in 1840 and proved to be invaluable in the war because of its "precise maneuverability, light draft, and towing power." *Nemesis* proved itself in battle many time over, including the conquest of the fortified city of Canton, normally safe from sea attack due to the dangers of the Pearl River to sailing warships. As Matzke explains, "*Nemesis* led the entire fleet to Canton itself on 21 May 1841 – an unprecedented feat – while carrying commanding officers and the men of the forty-ninth

31. Matzke, *Deterrence through Strength*, 37.
32. Matzke, *Deterrence through Strength*, 39-40.

regiment, and also towing a string of seventy to eighty boats with additional troops. Steam power let the British conquer a Chinese city that had been able to rely on its sea defenses for all of recorded history."[33] *Nemesis* and her sister steamships intimidated the Chinese so much they earned the name "demon ships."[34]

Nemesis is important because it encapsulates the uniqueness of British sea power during the Pax Britannica that space-power theory advises for space forces today. *Nemesis* was not a British government warship, but was combat power purchased for commercial reasons (namely pirate and bandit suppression) with private (or at least public-private partnership) funds, but commanded by a Royal Navy captain. *Nemesis* was also a pioneering technological marvel for her time for both her equipment and the tactics for steamship warfare she developed. *Nemesis* could not have been built without all of Great Britain's seapower and financial infrastructure contributing to her success. Moreover, of extreme importance to space-power theory, her transformational strategic technology was not in weapons, but in *movement and maneuver*.

Toward Dynamic Space Deterrence

With the true nature of the space domain understood, an outline of dynamic space deterrence based on space-power theory can be presented. The fundamentals of dynamic space deterrence strategy – deterrence through superior space control by rapid expansion of space capability across all elements of national power - are best explained through comparing them to nuclear-style static space deterrence elements when assessing their utility to the space domain.

33. Matzke, *Deterrence through Strength*, 137.
34. Matzke, *Deterrence through Strength*, 124-5.

The primary benefit that the defender has in a space attack is the problem of simultaneity. This is why a successful "first strike" against US space assets (potentially barring a nuclear electromagnetic pulse – EMP - attack, but even that is unlikely to be completely successful) would be almost impossibly difficult. The simultaneity problem was particularly important concerning Cold War nuclear strategy. The simultaneity problem is an issue that confronts a belligerent that wants to conduct a surprise "first strike" on the other's nuclear forces with their own nuclear forces. This first strike is meant to destroy all or at least a sizeable portion of the nuclear forces of the other side. Unless the first strike gets every single nuclear weapon with the potential to strike back, the attacker will suffer some nuclear damage. However, this is not the only problem that would confront an aggressor. Not only would they have to destroy all the defenders deterrence forces, they would also have to destroy them by surprise *at the same time*.

This is the simultaneity problem. If the Soviet Union took longer than, say, 30 minutes to destroy all US nuclear targets in the Cold War, the chances that some American forces would be alerted that the Soviet Union was conducting a massive attack and fire back before being destroyed would approach an almost absolute certainty. Therefore, the Soviet Union would have to destroy all American nuclear forces in a time short enough that no forces would be notified of an attack on the others before they too were destroyed.

Simultaneity makes the problem even more difficult for the Soviet Union than this because of the amount of activity and, consequently, tactical warning that would be available for the United States (who can see the preparations with their intelligence and reconnaissance capabilities) to know something was about to happen. Put simply, such a large attack would easily be seen coming days and perhaps weeks away. The demands in both surprise, quickness, and sheer force necessary for a completely successful nuclear "first strike" is simply incredible. This is why deliberate nuclear first strikes were considered far less likely than an "accidental" nuclear exchange in the second half of the Cold War when both US and USSR nuclear forces were developed with adequate detection, command and control, and rapid reaction striking force. Any first-strike, even if successful beyond reasonable expectation, would virtually assure unacceptable nuclear damage to the attacker in retaliation. First-strikes, put simply, did not work.

Simultaneity equally favors the defender in a space first-strike scenario. The sheer number of forces required to completely destroy the United States Space Force's constellations in space is enormous. Space-based nuclear detonations focused on generating electromagnetic pulse (EMP) waves to destroy spacecraft electronics may be able to conduct a successful first strike in space, but there would likely have to be multiple EMP weapons launched in conjunction to achieve victory, perhaps four or more. At these levels, their preparation would likely tip off defenders before the attack could take place. Then, space first strikes would begin to fall into the same traps that nuclear first strikes would need to overcome.

First, there would need to be an adequate number of EMP warheads mated to launchers, space-to-space attack satellites on-orbit, and ground-, sea-, and air-based weapons such as direct ascent ASATs and lasers to destroy all American space assets in one attack (which can be identified by intelligence with very long lead times, perhaps even years). Second, they would all have to be deployed in the appropriate locations from which to conduct the attack (which can easily be seen with long lead times, potentially months). Third, they would have to be targeted and ready to launch and destroy their targets immediately (which will have a lead time likely in days or weeks). Lastly, they will have to be targeted to destroy all of their targets in a very short amount of time (potentially hours or maybe even minutes) in order to secure an acceptably limited retaliatory strike in space or on Earth. In some ways, space first strikes would be even harder than nuclear first strikes due to the vast distances between space targets. Successfully hitting all targets in the span of a few minutes or hours may require weapons to be launched hours before the other strikes in order to hit all targets simultaneously, thus ensuring the defender has adequate warning of an attack to eliminate the first-strike's ability to surprise. A successful space first strike would be as prohibitively difficult to achieve as a nuclear first strike.

Since nuclear-style, blinding first-strike attacks on space assets are so difficult to achieve, they are very likely never to be attempted. Space wars would not necessarily be over in minutes or days until far more space weapons are placed in operation than currently exist. Therefore, with relatively few space-based weapons capable of participating in a first-strike blinding attack, any space warfare is likely to provide sufficient pause for defending forces to react and recover from the attack and give both belligerents sufficient time to impose political will well before automatic escalation can occur. Sufficient time to redeploy forces, deploy substitutes, and consider responses will drive space wars to look increasingly as limited warfare measures than automatic escalatory triggers that could potentially lead to nuclear warfare.

Concerns over a space EMP attack with a single nuclear weapon are perhaps the most legitimate of all "large-scale" space attack scenarios. The Nonproliferation Policy Education Center's (NPEC) January 2023 occasional paper "Pyongyang Goes Nuclear in Space: An After-Action Report" identified two potential scenarios where a nuclear EMP space attack might occur: as a prelude to a full-scale invasion of a neighbor with a high possibility of US involvement (i.e. China attacking US space assets prior to invading Taiwan) and, their adopted scenario, a rogue nation launching an EMP attack because they fear they are under immediate threat of regime change (North Korea pre-emptively attacking the US/ROK alliance for fear of being invaded.).[35] It is important to note that both envisioned scenarios involve vital national interests of the potential attackers, suggesting that even if space warfare is naturally a form of limited warfare, space assets and weapons can be used to threaten the vital interiors of a country either directly (in the form of space-to-ground weapons) or indirectly as enablers to joint force weapons that can threaten the vital interiors of a country.

With current technologies and deployed systems, such an attack may be difficult or impossible to intercept or defend against. Additionally, states whose vital national interests are sufficiently *actively* threatened will likely not be deterred from launching a space EMP attack. Thus, the key to preventing such an attack may be to embrace the limited warfare nature of space warfare and cease to threaten the vital interiors of countries with space forces entirely, whether in the form of weapons or combat support systems. Other deterrence measures such as encouraging potential adversaries to rely on their own space forces for national defense and development may make the costs of such an indiscriminate space attack prohibitively costly as well. The key to deterring single space EMP attacks may lie in assurances that a potential adversary's vital interiors (remember, China sees Taiwan as its own territory) is not a threat and to encourage all nations to develop their space infrastructures to make indiscriminate attacks less attractive to everyone – including potential adversaries.

However, even single space nuclear EMP attacks will fail the simultaneity problem. Single attacks will not be able to destroy all US space assets and even the most pessimistic of opinions on the viability of an EMP attack believes that many of the space system casualties of an EMP attack would be degraded by radiation over

35. Henry Sokolski (ed.), *Pyongyang Goes Nuclear in Space: An After Action Report*, Nonproliferation Policy Education Center (NPEC) Occasional Paper 2301, January 2023, 5-6.

time. Thus, a space-enabled joint force would still retain many of its space capabilities for at least some time and would likely be able to generate an aggressive retaliatory strike. A single space EMP attack would not blind and cripple the United States immediately, and the United States would likely respond to such an attack with overwhelming force. A space EMP attack would be neither easy nor cost-free. Combining this fact with an effort to ensure space forces are used primarily in defensive roles and not used to threaten the vital national interests of potential adversaries would go far to limit the threat of a space EMP attack.

If we begin to see space warfare as a type of limited warfare rather than an automatic escalation to nuclear conflict, many significantly different concepts of deterrence can begin to be imagined.

Static versus Dynamic Space Deterrence: A Balloon Analogy

A simple but effective analogy to understand static versus dynamic space deterrence might be to consider World War I airpower. Space forces in the past through today – and the thinking behind their use – can be considered a space equivalent to the "Balloon Mentality" of airpower. Satellites today are directly analogous in their lack of movement and maneuver capability to military balloons used for observation since at least the American Civil War. Balloons were true airpower and were militarily significant, but because they were incapable of relevant movement and maneuver, balloons could not form the basis of any useful military theory of airpower. It is difficult to imagine that an effective U.S. Air Force could be formed using the balloon, nor even the balloon's movement but not maneuver-capable cousin, the airship, as its central combat unit. Only an airplane can offer robust airpower.

In World War I, balloons spotted for artillery and aircraft and observed troop movements. Airships scouted for submarines and zeppelins bombed cities. Airplanes bombed targets, attacked troops, shot down balloons and airships, and fought for command of the air with other aircraft. But in the end, only airplanes had the capabilities to become truly capable military aircraft. Almost all WWI-era balloon and airship missions were quickly transferred to airplanes and balloons have disappeared from military inventories for all but the most ancillary and specific missions. Perhaps more importantly, there is no record of balloon or airship-on-airship engagements in airpower history. Only the airplane offered true offensive air capabilities. Balloon-based airpower was incapable of conducting many military missions, including fighting for command of the air. Only Airplane-based airpower and the mental change of the art of the possible that airplanes provided could develop true airpower.

Space forces, facing the same limited capabilities for movement and maneuver as WWI air forces, will probably follow a similar pattern. Satellites, sharing many of the attributes of the balloon, have military utility but are not really militarily-capable units. Our modern constellations of terrestrial support satellites are technological marvels, but they are not capable military vehicles, nor is any type of mobility-limited satellite likely to become one. But the US Space Force is built upon the satellite as its central unit. The United States suffers from a "Balloon Mentality" in space. Attempting to develop space warfighting doctrine and capability focused on the satellite is probably an impossible, yet astronomically costly, exercise in futility.

Satellites are not only likely incapable of effective self-defense, they are also very likely *incapable of effective offense* against other satellites, again mostly through their lack of movement and maneuver. Consider the Chinese Shijian-17 (SJ-17) and Shijian-21 (SJ-21) satellites. Many see these vehicles as space weapons.

However, their movement and maneuver capability against even quasi-position-bound satellites are not extremely significant. It is doubtful that these satellites can successfully attack more than one or two even minimally non-cooperative spacecraft in their operational lifetime, especially if an attacked spacecraft is commanded to "evade" or even "fight back" using its limited maneuver capabilities to resist. To be even partially threatening to US constellations, multiple SJ-17 and SJ-21-type satellites will need to be deployed across near-Earth space to place US constellations at significant risk of mission failure. Further, the time and effort required to position such an attacking satellite fleet correctly would make any such operation slow and easily detected, offering strategic warning and ample reaction time for the US to raise alerts and take defensive action. Satellite-on-satellite attacks will probably never rise to the level of a "bolt from the blue" rapid first strike. Like their balloon air-cousins, "Balloon Mentality" satellite engagements will probably be either completely non-existent or few and far between without any consequences of note.

Balloons did not fight balloons in WWI. Airplanes were the only active air threats to balloons. Airplanes also fought other airplanes to defend allied balloons and airships. Similarly, only an "Airplane Mentality" space force that places a premium on movement and maneuver forces will likely be able to offer a true space threat to satellites. Space weapons such as the SJ-17 and SJ-21 are as vulnerable to terrestrial-based direct-ascent ASAT missiles as are GPS satellites. Balloons will always be easy prey for airplanes. Movement and maneuver will always triumph, in any warfighting domain. "Balloon Mentality" satellites, even armed ones like the SJ-17 and SJ-21, will be incapable of defending themselves from

"Airplane Mentality" maneuverable space vehicles. In short, only a Starship* can offer robust space-power.

Space warfare solutions, even to defend the "Balloon Mentality" legacy space support constellations the US Space Force worries about defending today, will not be found in "Balloon Mentality" satellites, but rather a dynamic space "Airplane Mentality" supported by space vehicles with true movement and maneuver capabilities. New spacecraft such as the SpaceX Starship and other refuellable, high maneuverability vehicles (the US Space Force X-37B is a mere satellite with wings and should, at best, be considered a space zeppelin) or small, cheap and versatile microsatellites such as those being developed by the "New Space" commercial space companies championed by the Defense Innovation Unit's Space Portfolio will fight space wars, not movement and maneuver-restricted satellites confined forever to specific orbital planes.

It is better today, with the SpaceX Starship nearing orbital flight and true space movement and maneuver so close at hand, to embrace the "Airplane Mentality" of expansive space maneuver using true spacecraft for space deterrence and defense, rather than spend far more significant sums trying to defend our satellites under the "Balloon Mentality" under which many sanctuary-inspired theorists suffer.[36] Defense analysts should rather look to emerging "Airplane Mentality" space writers such as Jeffrey Becker.[37] It will be Airplane

---

* Large, high delta V, reusable maneuver spacecraft like the refuel-able SpaceX Starship.

36. For a very comprehensive and characteristic account of space warfare from such a "Balloon Mentality" perspective, see Brian Chow, "Stalkers in Space: Defeating the Threat," *Strategic Studies Quarterly* (Maxwell AFB, Alabama: Air University Press, Summer 2017), 82-116, https://www.airuniversity.af.edu/Portals/10/SSQ/documents/Volume-11_Issue-2/Chow.pdf

37. Jeffrey Becker, "A Starcruiser for the Space Force: Thinking Through the Imminent Transformation of Spacepower," *War on the Rocks,* 19 May 2021, https://warontherocks.com/2021/05/a-starcruiser-for-space-force-thinking-through-the-imminent-transformation-of-spacepower/

Mentality spacecraft the finally and conclusively expose the fallacies of the Sanctuary school and – perhaps paradoxically to some – prove that space is a defense-dominant domain with space forces as agents of limited war rather than the harbinger of nuclear doom.

Enhancing Dynamic Space Deterrence though the Joint Functions

Space-power theory-based dynamic space deterrence with an Airplane Mentality, then, should be focused on the credible threat of victory in space-to-space warfare through superior space control *only* in order to retain the space domain's strategic nature as a defense-dominant domain of limited war. Space forces capable of credible space deterrence rely on superior space-to-space combat ability to win tactical engagements while attacking enemy CLOCs and defending their own. Developing superior combat ability will rely on enhancing a space force's ability to successfully conduct the six joint functions in space. These functions are: command and control, fires, movement and maneuver, sustainment, and protection. All functions are required and improving space operations in any of these functions will advance space control, but some functions are more important than others.

Command and control (C2) is the ability of commanders to direct systems in combat and the responsiveness of those systems to those orders. Today, C2 nodes and their space systems are often hundreds or thousands of kilometers apart, with little tactical situational awareness. The belligerent that has better sensors on their spacecraft and faster C2 capabilities will generally win any engagements. Revolutions in space C2 beyond evolutionary improvements in traditional C2 constructs may be from artificial intelligence in the form of either "machine-aided" decision making or by allowing onboard computers to taken independent tactical actions in response to immediate threats. Another revolution may be the addition of crewed vehicles that will be able to take immediate action on-the-spot. C2 improvements could have a mild to moderate impact on space control relative to the other joint functions.

Intelligence, the ability to both know and comprehend what is happening in the space domain, is a critical function in space. Alternatively called space situational awareness or space domain awareness, sufficient intelligence is required in order to take any intelligent action in space. Enhancing understanding of adversary actions in the space domain for attribution and being able to detect and track space objects from the smallest piece of debris to the most active spacecraft, is always useful. However, improved intelligence is more important to strategic decision making than tactical warfighting. Furthermore, American space situational awareness is actually fairly advanced in relation to the other joint functions. Improved intelligence would have a relatively mild impact on space control relative to improvements in other joint functions.

Fires, or space-to-space weapons, is the ability to inflict force on a target object for reversible or irreversible disruption, damage, or destruction. Fires exists today in the form of reversible jamming, reversible or irreversible laser dazzling of sensitive sensors to disrupt or degrade spacecraft payloads, and destructive weapons such as antisatellite missiles and destructive directed energy weapons. Although such advances as directed energy lasers and particle beams or kinetic railguns and missiles or slugs can certainly improve space control, today fires is not the most critical deficiency in modern space forces. With sufficient velocity, anything can become a weapon. Improvements in fires will likely only have a mild- to- moderate impact on space control.

Movement and maneuver, the ability to move positions and maneuver to tactical advantage, in space terminology – propulsion, is by far the most critical joint function to improve in space control. Movement and maneuver is the key joint function to allow the defense dominant nature of the space domain take full effect on space operations. Improved movement through advanced space propulsion will allow space systems to be more strategically, operationally, and tactically relevant across all space missions. Movement-enabled space systems can be more responsive to joint requirements. Maneuver will allow better defense by offering the ability to dodge incoming attacks. Rapid space lift, such as may be provided by SpaceX Starship systems, could move rapidly with modular "emulator payloads" to replace the payload coverage of destroyed satellites. Defensive spacecraft with advanced propulsion could better respond to and outmaneuver attacking spacecraft and eliminating attackers before they reach their targets. Movement and maneuver, whether increasing the reaction-capability of satellite constellations or breathing life into dedicated protector or reconstitution systems, will add immensely to space control.

Sustainment is intimately connected to movement and maneuver. Rapid movement and maneuver in space requires immense energies, and additional propellant and the ability to refuel spacecraft propulsion systems will become every bit as important a capability as refueling is in the air, land, and sea. Rapid strategic movement and tactical maneuver depend on sustainment capabilities. Sustainment capabilities will likely require "positional" orbital fuel depots as well as "responsive" space refueling craft like naval oilers or jet tankers. Sustainment will need to improve *before* any significant movement and maneuver advances will be possible, and it will be one of the most important functions to improve to enhance space control.

Finally, protection is the ability for a system to defend itself and absorb damage. Protection is perhaps the most neglected joint function in space due to the high cost of space lift and, consequently, the design requirement for spacecraft to be as limited in mass as possible. Protective systems have been ignored in favor of cost and payload considerations. Advances in protection for current spacecraft could include "escort" spacecraft that can intercept potential attackers or deploy large "umbrella" systems to hide the spacecraft its detecting from tracking or weapons targeting while over certain areas. Ultimately, as the cost of space lift decreases and large-scale construction on orbit emerges, protection in the form of active defensive weapons and even armor of various character will allow the space domain's defense-dominant nature to bloom in full form. Protection will have a very high impact to space control relative to other joint functions, but likely only after movement and maneuver makes larger and heavier space systems economically feasible.

It must be noted that improving any and all of the joint functions in space will ever increase the space domain's defense-dominant characteristics. Seemingly paradoxically, the more man learns to fight in space, the less danger space warfare will present to Earth. As military historian Martin Van Creveld says, "it is only when man uses an environment for war that he really comes to dominate it."[38]

A Warning against Space-to-Ground Weapons

An important issue to space deterrence is the type of space weapons that are fielded and deployed. Space weapons can be permanent (destructive directed energy or kinetic weapons) or reversible (electromagnetic jammers or dazzlers that return the target to complete functionality after turned off). Reversible space-to-space weapons are generally preferred because they are, by nature, non-destructive temporary "mission kill" systems rather than "hard kill"

38. Martin Van Creveld, *A History of Strategy: From Sun Tzu to William S. Lind* (Kouvola, Finland: Castalia House, 2015), 123.

systems and the target suffers no permanent damage to their space systems. Reversible weapons work remarkably well in the "limited war" space conflict paradigm.

Destructive space weapons (deployed in space) are more complicated. These weapons can be further divided into two main groups, space-to-space (target enemy space vehicles from orbit) and space-to-ground (target terrestrial objects from space such as aircraft, naval ships, and ground position) weapons. Since space-power theory has tended to converge that space warfare is a centered on the defense of friendly CLOCs and attack against adversary CLOCs, space-to-space (S2S) weapons have a clear mission in space warfare. S2S weapons systems are meant to defend and attack CLOCs. S2S weapons therefore conform to the nature of space warfare. They can be used an attempt at a "nuclear first strike-like" attack on an adversary's entire space infrastructure in the "escalation" framework or in the limited space war paradigm. However, they have special place in the limited war framework because they are intended to contest CLOCs, not population centers or fielded forces, and thus intend to inflict economic damage only. Any loss of life from an attack against a CLOC would likely be nonexistent or minimal compared to terrestrial fighting and would not be theoretically necessary at all and therefore would be incidental, not intended. Attacks against CLOCs may be conducted in service to a first strike but are by nature themselves expressions of limited war.

Space-to-ground (S2G) weapons are by no means expressions of limited war. Although Dolman offers that S2G weapons may be necessary to deter or defeat terrestrial anti-satellite threats[39], they are not as clearly supported by space-power theory as S2S weapons because they have little connection with controlling CLOCs. There

---

39. Everett C. Dolman, *Astropolitik: Classical Geopolitics in the Space Age* (London, United Kingdom: Routledge, 2002), 157.

is no doubt that attacks on space launch facilities and other terrestrial infrastructure can severely impair a nation's space efforts, but permanent orbital S2G weapons would be easily identified for their size, magazine, and power requirements. Therefore, they can be alleviated somewhat by terrestrial maneuver of potential targets out of striking range from S2G orbital paths. They will also likely be prime targets for S2S attack systems that would likely be far cheaper than their S2G targets, and much less likely to survive the encounter. S2G systems may not make economic and military sense.

However, the real problem with S2G weapons is that they are inherently escalatory and do not comport to a limited war space paradigm. The most likely S2G weapons to be deployed (indeed, many such systems have already been developed and tested) are Fractional Orbit Bombardment Systems (FOBS)-like systems that only orbit for a short time before dropping on an enemy's territory. These types of weapons, much like hypersonic cruise missiles and permanently orbital S2G weapons, are designed to violate the territorial integrity of nations. They place the entire territory of an adversary's nation at risk, and with the capability of hosting nuclear payloads, any strike into the interior of a nation may be considered a potential nuclear attack, fundamentally altering a space deterrence problem immediately into a nuclear first strike problem, which might well begin an inevitable escalation to general nuclear war.

If the world wants to achieve a stable multipolar world with a limited war space paradigm that significantly reduces the chances of wide-scale general conflict and escalation, deploying S2G weapons must be avoided. The limited war space paradigm is predicated upon great powers being confident that their territorial integrity as regional hegemons is secure. Instead of supporting territorial integrity, S2G weapons can threaten all of a nation's interior with up to nuclear force. S2G weapons trade the potential security of a limited war space paradigm for an inherently escalatory and unstable nuclear dilemma for no perceivable benefit or support from space-power theory. S2G weapons are against the nature of the space domain, according to space-power theory. Although S2G weapons may have some utility in defending space systems from an attack from terrestrial antisatellite weapons, the "juice is probably not worth the squeeze." Defense against terrestrial anti-satellite weapons may be better found in the nature of the space domain itself, which will be described later.

It is important for space enthusiasts to resist S2G weapons because the United States Space Force's incompatible airpower strategic heritage links the ability to direct force on the terrestrial battlefield with service independence legitimacy, as argued by the Air Force Association when lobbying against the Space Force's establishment in 2019.[40] It is very likely many Space Force personnel believes that S2G weapons are highly desirable. Deployment of S2G weapons by any great power in significant operational number runs the extreme risk of permanently wedding nuclear escalation to military space activity to the world's detriment.

A Vision of Peaceful Space Expansion but Limited Space Warfare

Although many agree that the world is transitioning from a unipolar to multipolar order, this system clearly has not yet fully emerged. A stable multipolar order (that may or may not be in the United States' interest) may emerge if the aspiring regional hegemon China finds some accommodation that satisfies their concerns about Taiwan and the South China Sea and if the war in Ukraine ends with a stable

---

40. Michael Martindale and David Deptula, "Organizing Spacepower: Conditions for Creating a US Space Force," Mitchell Institute Policy Paper Vol 16, August 2018. a2dd91_2ff8dfe95e694f80b4139d05650843ed.pdf (mitchellaerospacepower.org)

border mildly acceptable to Russia and Europe. If the military "Great Powers" – the United States, China, Russia, India, and perhaps the European Union – each become dominant in their respective regions and no power or alliance of powers can meaningfully project military power in the regions of any other power, the importance of space systems for power projection warfighting will diminish significantly. In a world of stable regional hegemons generally secure in their territorial integrity, the requirements for space support to terrestrial militaries decrease significantly. Communications, navigation, and intelligence, surveillance, and reconnaissance (ISR) needs diminish greatly for garrison activities and if large military formations are not sent abroad thousands of miles from home territory, these services are much more easily accessed from terrestrial sources, and space sources are much more easily defended. Instead of being extremely valuable and presaging a nuclear first strike if attacked, terrestrial support space forces will lose much of their immediacy and space attacks will go from the feared "first strike" scenario to a more measured, less violent, normalization into a "limited war-type" space combat paradigm where space attacks are generally seen as a non- or limited-escalation method of political signaling or mildly harassing maneuvers – much like today's multitude of limited space jamming efforts across the globe that is mostly ignored, rather than major world-changing events.

Moving to a "limited space-centric" space war paradigm will be extremely important to the safety and security of all nations, especially their territorial integrity. Thus, even if military space activity increases dramatically, the larger interests of deterring great power war on Earth would be served exceptionally well. Safe from other great power threat through natural barriers such as oceans or ethnic terrain and proliferated space systems that can provide adequate warning of impending attack to make both nuclear and conventional deterrence of terrestrial warfare viable for all great powers, and blocked from terrestrial expansion by equally formidable barriers, great powers will have little choice that to move great power competition to economic and social development. The rich bounty of space resources will be one of the primary drivers of economic development in the twenty-first century.

Instead of competing over or fighting for control of dwindling resources on Earth, powers that are secure in their territorial integrity and unable to gain extra territory by conquest due to the supremacy of defense in a multipolar world will choose to leverage the vibrant commercial space industry to reach for the abundant resources available in space to satisfy their need for resources. Great powers will shift state resources from terrestrial militaries to space activities to develop profitable space industries and resource production facilities, including space forces to protect their access to them. Space forces will begin to defend CLOCs that transmit necessary resources and information critical to national development. In time, space will be the primary lever of the balance of power on Earth as established great powers – and emerging middle powers intending to become great through space activity like Portugal or Great Britain in the Age of Discovery – send their best, brightest, and most energetic to conquer outer space and harness space's vast riches. This new and permanent space race will quickly take on the feel of the European Age of Discovery with its centrality of naval power in the world order, with new great space powers with space navies potentially eclipsing traditional great powers.

In this new space age, space wars will not be the catastrophic prelude to an irreversible escalation to all-out nuclear conflict that Sanctuary advocates fear. Rather, any space wars will generally take on the form of the "quasi-wars" of the 18th and 19th centuries that focused on naval warfare in service to economic competition. The space infrastructures of even medium space powers would be highly resistant to large-scale terrestrial attack because the defense dominates in the space domain with a sufficient mastery of space control. Since large-scale space warfare would be extremely difficult, expensive, and risky, regular space warfare will focus on gaining small advances in economic or commercial advantage from the space domain. Any wars that would be fought would be primarily limited to space (for Earth wars would be unprofitable due to the strength of defense in a stable multipolar order of regional hegemons) and involve few lives at risk with fewer overall casualties than terrestrial war.

Furthermore, war would generally not be a common space event. Since few resources exist in low Earth orbit (LEO) and the orbital altitudes close to Earth would be filled with critical infrastructural elements such as communications (such as Starlink internet constellation), intelligence (such as national systems and commercial constellations like Planet Earth observation satellites), LEO combat would be exceptionally rare. Especially because space systems would generally not allow for global power projection, they would be of little value to target by any adversary. Global navigation systems such as GPS, BeiDou, Galileo, and GLONASS in medium Earth orbits (MEO) would be critical to the global economy and relatively insignificant military targets in equal measure, making them unlikely to be targeted. Geosynchronous orbits (GEO) would also be filled with critical systems with limited military target value. GEO communications satellites will be augmented or even supplanted by constellations in LEO and the requirements of large bandwidths to support large armies deployed away from host countries will be significantly reduced. Large and economically significant platforms in GEO such as space-based solar power platforms supplying clean energy for Earth will likely be capable of defending themselves well from terrestrial or space attack due to the defensive dominant domain gravity well.

It cannot be stressed enough that if global power projection by one or a coalition of great powers that can threaten the territorial integrity of another is generally impossible, then the criticality of targeting space warfighting support systems diminishes rapidly. It is highly likely that in such an environment, a general drive toward banning or limiting space weapons systems in Earth orbit may begin in earnest. Military space operations, beyond "coast guard-like" support operations including search and rescue, general security and patrolling, aids to navigation, and space traffic management and anti-piracy (be it defending the electromagnetic spectrum from illegal jamming or the more conventional form of protecting against space theft on the celestial lines of communication), may generally become pointless.

In such a world, limited skirmishes in deep space over large or important resource zones will probably be the most common military conflict points. Harassment of adversaries' CLOCs that may or may not be violent would probably be the most common active military aggression. But active violent military activity will probably be a minor activity at most compared to peaceful strategic offensives in space.

Peaceful strategic offensives in space (PSOs) are an expression of the dictum, adapted from Mahan, that space strategy is meant to expand a nation's space-power across all dimensions (diplomatic, information military, and economic) in peace as well as in war. Space strategy is active and integrated across the entire competition continuum. When not in armed conflict, where military space activity is preoccupied with defending friendly CLOCs and attacking the CLOC's of the enemy, the primary activity for an expanding space power is to strengthen and expand existing CLOCs and build new ones to increase one's informational and economic space control.

PSOs are activities conducted primarily in times of general cooperation or competition below armed conflict in the competition continuum to expand the nation's general space control capabilities (be it scientific, economic, or military control) through construction or enhancement of national CLOCs. These PSOs can take the form of exploring and assaying new regions of space for resources, conducting basic and applied research on the space domain and technologies that can increase the ability to operate effectively in it, and – most importantly – to develop new profitable industries and resources generators that become profitable businesses and positive economic activity. Economic space-power, especially in the form of profitable commercial enterprises, is the cornerstone of effective and sustainable national space programs. While corporate or scientific agents will likely lead, manage, and operate the most important parts of PSOs, space militaries will also play a role. Whether trade follows the flag or the flag follows trade, what can be certain is that the flag and trade will almost always be paired closely to ensure the Space Virtuous Cycle is allowed to benefit the nation as a whole. Militaries will provide defense from human and natural threats as well as secure the property and lives that comprise friendly CLOCs. It is in patrolling CLOCS that space forces will likely find the most friction points with which violence might erupt into space conflict. The farther away from Earth these conflicts take place, the more likely they will resemble limited naval commercial wars than truly destructive terrestrial world wars.

Again, in this world the continued effort to improving mastery of the joint functions in space is critical to space superiority. Lessons learned and technologies developed from PSOs will improve space forces' abilities to conduct all joint functions in space, thus continuing to increase the defense-dominant character of the space domain as historical technical shackles limiting human space activity are eliminated. As all nations improve their ability to operate in space, both general deterrence and deterrence of major wars in space will be improved considerably – especially lowering the risk of military space activity escalating into general conventional or nuclear war on Earth.

If we briefly return to the scenario of a rogue state launching a single nuclear EMP attack at space systems we can see the power of mastering the joint functions in space. There are two damage mechanisms to this type of attack. The first is immediate line-of-sight EMP that can damage or destroy a space vehicle's electronics. The second is the potential for a nuclear blast to "energize" radiation belts in occupied orbits with charged particles that can degrade space vehicles over time. The damage potential of the nuclear blast itself is almost inconsequential. The destructive potential of EMP is based primarily on the fragility of current space systems. If a space power has devoted itself to mastering the joint functions in space, the severity of the EMP attack is greatly diminished. First, the space domain itself is bathed in radiation. Thus, conquering the space environment itself will likely improve radiation shielding and hardening of systems dramatically to the point that a single relatively low-yield nuclear explosion in orbits may not generate enough EMP to damage electronics immediately or charged particles to degrade modern systems over time. Therefore, enhancing protection against radiation will better defend constellations from this type of attack.

Furthermore, enhancing space system movement and maneuver further limits the threat of EMP attack. If a state launches a satellite that may be a nuclear EMP weapon, maneuverable systems can immediately be diverted to inspect the suspect vehicle. Improved intelligence collection technologies such as nuclear detection sensors could make positively identifying a nuclear payload much easier. Also, improved fires in the form of space-to-space (S2S) weapons can allow defensive systems to destroy or disable the EMP weapon before it can be deployed or used. If the EMP weapon is successfully fired, space systems with enhanced movement and maneuver can move to avoid high radiation areas, thus limiting their exposure to captured particle radiation. Virtually all joint functions can be employed to mitigate the risk of a rogue EMP attack. Ultimately, the danger of the rogue single EMP scenario is another function of the limited space technology available to today's spacefaring powers. By committing to develop advanced space technologies, nations can increasingly mitigate the risk and consequences of such an attack, and further moving space warfare into its limited war nature.

By adhering to a policy of dynamic deterrence in space through peaceful strategic offensives to increase space control, general conflict in Earth orbit can be addressed through the twin avenues of maximizing the natural (yet still hidden) truth that space is a defense – not offense - dominant domain and to minimize the attractiveness of Earth orbit space platforms as lucrative targets of a terrestrial nuclear first strike. More importantly, dynamic space deterrence will support, rather than stymie, rapid scientific, commercial, and even military expansion into space. By basing space deterrence on the nature of the space domain as expressed through space-power theory rather than the stale and incompatible assumptions of nuclear deterrence theory, the world's leading space nations can accrue ever more benefits of space for humanity while making the world (and beyond) an ever more peaceful domain.

Should Achieving Space Superiority Be America's Immediate Military Objective?

Peter L. Hays[41]

For millennia, the heavens above have fascinated and awed humanity as the dwelling place of God, or simply as a domain of unimaginable expanse and mystery. With the opening of the Space Age, concerns about space and security have become inexorably and increasingly intertwined with terrestrial politics while simultaneously intensifying many of humanity's most utopian dreams and dystopian fears. Humanity aspires to a hopeful future in space and needs a clear vision of the best options and pathways along which we can continue exploring and using space, because these approaches are likely to significantly shape our destiny for better or worse. Paradoxically, however, the best approach toward achieving a brighter long-term future in space must at present focus on the short-term and security dynamics in Earth's orbital space. Space security cannot be divorced from terrestrial security issues that humanity must address to manifest a brighter future both on Earth and in space.

As the foremost spacefaring state, the United States must play a leadership role in guiding humanity's interactions with the cosmos and finding ways to use space to enhance stability and security. This

41. The views expressed in this paper are those of the author and do not reflect the official guidance or position of George Washington University, the United States Government, the Department of Defense, the United States Air Force, or the United States Space Force.

paper argues that the United States should continue seeking space superiority to lead both on Earth and in further exploring space and harvesting space resources. It explains what space superiority is, why the United States needs it, and outlines some of the best approaches for pursuing such superiority. There are many domestic and international disagreements, constraints, and conflicts regarding the proper role for the United States in space; seeking space superiority provides the clearest and surest path forward to overcome these challenges.

What Is Space Superiority?

Space superiority is a complex concept shaped by the expanse of space as well as the attributes of orbital dynamics, space technologies, space operations, security dynamics, and military doctrine. In the most basic terms, space superiority means being able to do what you want in space, when and where you want to do it, as well as having ability to deny others from interfering with this activity. The authoritative joint doctrine for space operations defines space superiority as "the degree of control in space of one force over any others that permits the conduct of its operations at a given time and place without prohibitive interference from terrestrial and space-based threats."[42] This doctrine statement also indicates that the United States Space Force (USSF) "conducts space operations to achieve space superiority for the United States as a Service core competency."[43] Charged with achieving space superiority as a core competency, the USSF must prioritize ways and means of achieving this objective and determine how space superiority relates to its mission to support and enable Joint Force operations. In addition,

---

42. Joint Publication 3-14, "Space Operations," (Washington: Department of Defense, Ch 1, 26 October 2020), xiii.
43. Ibid., I-4.

the December 2021 Space Priorities Framework indicates that "access to and use of space is a vital national interest,"[44] the December 2020 National Space Policy stated that "unfettered access to, and freedom to operate in, space will remain a vital national interest,"[45] and the June 2020 Defense Space Strategy emphasized that the Department of Defense (DoD) "is taking innovative and bold actions to ensure space superiority and to secure the Nation's vital interests in space now and in the future."[46] It is clear that the United States is increasingly focused on understanding the requirements for pursuing space superiority and faces growing concerns about its ability to achieve and maintain this superiority.

The U.S. military uses the broader concept of "space control" and the phrase "freedom of action" to help describe and explain why space superiority is needed: "Space control includes offensive space control and defensive space control operations to ensure freedom of action in space and, when directed, defeat efforts to interfere with or attack US or allied space systems."[47]

> The purpose and value of space superiority is to provide the freedom of action in space in the pursuit and defense of national security interests. The US ability to capitalize on and protect space systems, and to counter enemy capabilities, contributes to US space superiority. To establish and maintain space superiority, commanders require resilient space forces that have the skill and the experience to protect and defend space systems across the competition continuum and to deny the same to the enemy. Commanders should

---

44. United States Space Priorities Framework (Washington: The White House, December 2021), 3.
45. National Space Policy of the United States of America (Washington: The White House, 9 December 2020), 1.
46. Defense Space Strategy Summary (Washington: Department of Defense, June 2020), 1. Vital national interests are generally considered to be those interests the United States would use military force to defend or pursue.
47. Joint Publication 3-14., xi.

> understand how to request space forces and determine whether the capabilities are continuously available or know when special authorization and coordination are required. The use of offensive and defensive operations in multiple portions of the OE [operational environment] may be necessary to maintain space superiority. To maintain space superiority, joint forces must have SSA [space situational awareness] and the knowledge, training, resources, and authorities needed to defend our ability to use space. This includes being prepared to prevent the threat from exploiting space against the United States and its allies and partners. Commander, United States Space Command (CDRUSSPACECOM), is responsible for establishing space superiority and may request specific contributions from other CCDRs [combatant commanders].[48]

Orbital dynamics and current space technology set important conditions and limitations for achieving space superiority. Despite the many visions from science fiction and fantasy that may overly influence our thinking about space, current technology dictates a much narrower view with an Earth-centric focus. Today, almost every satellite that was designed for or could support national security operates in one of four orbits of primary concern: low Earth orbit (LEO), highly elliptical orbit, medium Earth orbit, or geostationary Earth orbit (GEO). While cislunar space, the Moon, asteroids, and Mars hold much potential and inspire great longer-term ambitions,[49] the emphasis for our current and near-term space security analysis must be on the region that is bounded by GEO because operations in this region directly influence security on Earth. This much smaller, but still vast, region can be considered as

---

48. Ibid., I-4.
49. *National Cislunar Science & Technology Strategy* (Washington: National Science and Technology Council, November 2022).

the littoral or "cosmic coastline;"[50] understanding security dynamics in Earth's littoral region and how they shape terrestrial military operations and the balance of power on Earth provide an essential foundation for developing space superiority concepts.

Space provides unique positional advantages, persistent emplacement, and a global perspective, but great energy is required to move a satellite above the atmosphere at the bottom of Earth's gravity well and to accelerate it so it can sustain the specific orbit for which it was designed. Reaching higher orbits requires more energy and once established, significantly changing orbits also requires considerable energy. Accordingly, it is very rare for a satellite to maneuver away from one of the four main orbit types into which it was originally placed; satellites normally travel along highly predictable paths that vary little between the beginning and end of their operation.[51] These attributes of orbital dynamics, along with the lack of traditional cover and concealment measures available on Earth, means satellites can be more easily detected, tracked, and targeted than can terrestrial forces, which are routinely able to maneuver and hide. Because propagation tools can quite accurately predict where satellites will be days in advance, satellites are best considered to be like those known terrestrial military facilities or systems that could be moved but which rarely are.

---

50. The term cosmic coastline, considering orbital space as Earth's littoral region, and use of continental maritime theorists are major themes in Bleddyn Bowen, *War in Space: Strategy, Spacepower, Geopolitics* (Edinburgh: Edinburgh University Press, 2020). Michael V. "Coyote" Smith dismisses the Air Force's antiquated "aerospace continuum" concept by emphasizing the 65-mile-wide transverse region that separates the ceiling of aerodynamic flight from the floor of orbital flight and noting that sustained operations cannot be conducted within this region; Smith, "Ten Propositions Regarding Spacepower," (Maxwell AFB: Air University Press, October 2002).
51. Excellent sources on basic spacecraft designs and orbital dynamics are Jerry Jon Sellers, *Understanding Space: An Introduction to Astronautics* (New York: McGraw-Hill Primis, 2000) and *AU-18: Space Primer* (Maxwell AFB: Air University Press, 2012).

Attributes of space launch and orbital dynamics also drive space technology and operations in significant ways. Satellites are normally designed to be as light and small as possible to facilitate launch; they may unfurl and reconfigure into various asymmetrical configurations of solar arrays and antennas once in orbit outside atmospheric drag. Satellites must be made sturdy enough to withstand the rigors of launch and most have at least some hardening against naturally occurring radiation in space, but they almost never include the kind of protective materials found on most terrestrial military systems because of the weight of these protective measures. Traditional satellite architectures have been driven by several factors including the costs and dangers of space launch (still the most hazardous part of satellite operations), significant limitations on capability to service satellites, perceived economies from custom building very small numbers of increasingly capable and large satellites, and the ability of just a few of these highly capable satellites to perform a variety of key missions very competently. Due to these factors, several states, and the United States in particular, in the past chose to develop and operate a very small number of highly expensive, sophisticated, and exquisitely capable satellites. Each of these attributes add to the vulnerability of satellites and exacerbate temptations for enemies to negate them because they are so fragile, so few, increasingly important, operate in highly predictable ways, and cannot today be repaired on orbit.

Another important defining characteristic of almost all space systems is that they are dual use, meaning that they can be used for both civilian and military applications. This dual use characteristic has been inherent since the earliest days of space technology development and is highlighted by describing Wernher von Braun, the leading space technology pioneer, as a "dreamer of space, engineer of war."[52] Dual use considerations create difficult balancing and tradeoff issues as the United States and others attempt to promote space technologies and activities considered to be benign, while limiting similar things that may be threatening. In space, an implication of dual use is that any satellite that can transmit or maneuver could be used as a rudimentary anti-satellite (ASAT) weapon. Such use is not likely to be as effective as a purpose-built ASAT, but increasing development of in-orbit servicing and manufacturing (IOAM) and active debris removal (ADR) capabilities threatens to blur and complicate distinctions between commercial, civil, and military applications and operations. Similarities between ASAT systems and some of the technologies and operations of IOAM and ADR systems are so great that analysts worry that widespread development of these beneficial commercial and civil capabilities would also create significant but latent ASAT potential.[53]

Terrestrially, the U.S. Global Positioning System (GPS) provides an excellent illustration of dual use. GPS was originally designed as a military system, primarily to support employment of nuclear weapons, and its most accurate signals were to be reserved for military use only. Creation of a worldwide, multibillion dollar industry was enabled after the United States decided in 2000 to discontinue Selective Availability, the deliberate degradation of GPS signals for civilian use, and today the U.S. military comprises just a

---

52. Michael J. Neufeld, *Von Braun: Dreamer of Space Engineer of War* (New York: Vintage, 2008) is the authoritative assessment of von Braun's contributions and legacy.
53. See, for example, Brian G. Chow, "Space Arms Control: A Hybrid Approach," *Strategic Studies Quarterly* (Summer 2018): 107-132; and James Alver, Andrew Garza, and Christopher May, "An Analysis of the Potential Misuse of Active Debris Removal, On-Orbit Servicing, and Rendezvous & Proximity Operations Technologies," Capstone Paper, George Washington University, 6 May 2019.

small fraction of GPS users worldwide.[54] The GPS timing signal can be considered as a "global digital heartbeat" and is probably the most important space capability, because it provides a "handshake" that currently synchronizes connections between all global digital telecommunications networks, supports enhanced emergency location services, and delivers a location timestamp that facilitates many global financial transactions. The dual use characteristic of satellites is perhaps the single largest factor that complicates space security considerations, making it more difficult for analysts to determine ways to incentivize desired applications, constrain malign potential, and consider how these factors shape space superiority.

Dedicated or dual use capabilities can strengthen security, but security is an ambiguous and relative concept. Analysts use the term "security dilemma" to describe these relative and interactive aspects of security and study them as a cause of war and one of the central problems of international relations.[55] The characteristics of satellites exacerbate some of these issues and make their relative contributions to security more ambiguous and elusive. More specific characteristics of satellites contribute to the ways they may promote or inhibit cooperation under the security dilemma. In a seminal 1978 article, Robert Jervis applied game theory approaches to scenarios commonly used to analyze causes of war such as Stag Hunt and Prisoner's Dilemma, positing that two variables are primary determinants of how likely or unlikely it is that states can achieve cooperation: 1) whether offensive or defensive capabilities have the advantage; and 2) whether analysts can distinguish between

---

54. On Selective Availability, see https://gps.gov/systems/gps/modernization/sa. A 1995 RAND Study led by Scott Pace, *The Global Positioning System: Assessing National Policies*, was instrumental in moving the United States toward discontinuing Selective Availability.
55. Some of the earliest and most influential analyses of the relative and ambiguous characteristics of security include John Herz, "Idealist Internationalism and the Security Dilemma," *World Politics* (January 1950) 157-180; and Arnold Wolfers, "'National Security' as an Ambiguous Symbol," *Political Science Quarterly* (December 1952): 481-502.

offensive and defensive capabilities.[56] Applying this framework creates a 2x2 matrix in which Jervis labels situations where offense has the advantage and analysts cannot distinguish between offensive and defensive capabilities as "doubly dangerous" and situations with the opposite conditions as "doubly stable."[57] Some disagree, but unfortunately today most analysts perceive that the doubly dangerous situation corresponds most closely to the current characteristics of space. Jervis finds that this situation "is the worst for status-quo states. There is no way to get security without menacing others, and security through defense is terribly difficult to obtain."[58]

Brad Townsend builds from Jervis' analysis; extends it by incorporating work from Stephen Van Evera, Charles Glasser, and Evan Montgomery; and applies it directly to space, finding that the current space security situation is less dire than Jervis' original formulation would predict.[59] On the question whether offensive or defensive forces have the advantage, Townsend agrees with most analysts that at the tactical level of war the offense has strong advantages when attacking individual satellites, but he finds that at the strategic level the defense has slight advantages in countering attacks against proliferated and robust satellite architectures. On the question whether analysts can distinguish between offensive and defensive capabilities, Townsend notes that peaceful or nonaggressive uses of space have become nearly synonymous with defensive uses; however, because conventional forces have become so dependent on these capabilities for both defensive and offensive

---

56. Robert Jervis, "Cooperation Under the Security Dilemma," *World Politics* (January 1978): 167-214.
57. Ibid., 211.
58. Ibid.
59. See Brad Townsend, "Strategic Choice and the Orbital Security Dilemma," *Strategic Studies Quarterly* (Spring 2020): 64-90; and *Security and Stability in the New Space Age: The Orbital Security Dilemma* (Milton Park, UK: Taylor & Francis, 2020).

terrestrial military operations, these ostensibly defensive capabilities have become tempting targets that create strong incentives for development of counterspace capabilities. Nonetheless, he posits that "distinguishability remains relatively high among existing space systems."[60] We will return to these important points from Jervis and Townsend below when discussing specific actions the United States can take to achieve and maintain space superiority.

Legal and governance structures form another set of foundational factors that some believe set limitations on concepts related to space superiority. The Outer Space Treaty of 1967 and subsequent agreements on rescue and return of astronauts; liability from terrestrial damages caused by space launch, reentry, or during space operations; and registration of space objects provide a quite broad but not deep or fully developed legal foundation. These agreements set out lofty aspirational principles, while codifying only minimal constraints and procedures that were acceptable to the superpowers – the major space actors at the time the agreements were instituted. The legal regime was not initially designed for comprehensive governance over a fuller suite of concerns including overall security, debris, and commercial issues; it has grown increasingly inadequate to address current and future concerns as problems multiply and more and increasingly diverse space stakeholders are added. Space has been militarized since the beginning of space operations and the legal regime does little to preclude the weaponization of space except for nuclear or weapons of mass destruction. In short, the legal regime is aspirational, there is no current international appetite for relitigating foundational principles or reopening negotiations, and it has become less relevant as space militarization continues, military confrontations increase, and the potential for space weaponization increases. This regime simply cannot govern whether and how states

60. Townsend, "Strategic Choice," 84.

may pursue space superiority unless it is significantly enhanced, made more specific and binding, and robust compliance verification measures are developed.

As illustrated by the stunning successes of the Ukrainians in defending their country following the Russian invasion, the value of commercial space systems in supporting a wide range of military operations has grown exponentially.[61] These commercial capabilities provide critical information that strengthens worldwide support for Ukraine, supply communications connectivity that helps coordinate many Ukrainian military operations, and demonstrate that states do not necessarily need to own and operate space systems to use them effectively. This changed context requires the United States to accelerate assessments about the ways commercial systems contribute to space superiority and strengthen measures to protect and deter attacks on commercial space systems supporting national security. Overall, to better leverage enhanced development of commercial space capabilities, the USSF should first comprehensively, transparently, and consistently determine those space mission areas where it can rely on commercial services or international partners. Some space mission areas such as nuclear command, control, and communications should never be outsourced, while others such as environmental monitoring might be completely provided by commercial or international partners. Other mission areas could be mixed between U.S. or allied operated and commercially provided. A whole-of-government approach is then needed to assess the commercial viability of those services upon which the U.S. government intends to rely, either wholly or in part, and the government should act to improve the commercial

61. See, for example, Benjamin Schmitt, "The Sky's Not the Limit: Space Aid to Ukraine," Center for European Policy Analysis, 19 May 2022; David T. Burbach, "Early Lessons from the Russia-Ukraine war as a space conflict," Atlantic Council, 30 August 2022; and Jonathan Beale, "Space, the unseen frontier in the war in Ukraine," BBC News, 6 October 2022.

viability of these services. This is not explicitly picking winners, but rather using comprehensive approaches such as serving as an anchor tenant, structuring import-export bank loans, adjusting export controls, and tailoring licensing and regulatory procedures to improve the commercial viability of commercial space services that the U.S. Government has determined that it will rely upon and contribute to space superiority.

The United States should also continue supporting and advancing international approaches to strengthen deterrence of attacks on commercial space capabilities and improve protection measures for these systems. Helpful steps include the United Nations Committee on the Peaceful Uses of Outer Space promulgation of 21 guidelines for the Long-term Sustainability of Space Activities,[62] the North Atlantic Treaty Organization's declaration that space is a fifth operational domain where attacks could invoke Article 5 defense obligations,[63] the DoD's tenets of responsible space behavior in space,[64] and the pledge by the United States that it will no longer conduct destructive direct ascent ASAT tests that has now been joined by several other states and adopted by the United Nations General Assembly.[65] These steps and others seem to be generating

62. Report of the Committee on the Peaceful Uses of Outer Space, Sixty-second session (12-21 June 2019), 54-69.
63. NATO's decision to consider space an operational domain like land, sea, air, and cyber is helpful but it added caveats weakening Article 5 obligations for attacks in space: "A decision as to when such attacks would lead to the invocation of Article 5 would be taken by the North Atlantic Council on a case-by-case basis." See "NATO's Overarching Space Policy." NATO, 17 January 2022.
64. Secretary of Defense, "Tenets of Responsible Behavior in Space," (Washington: Department of Defense, 7 July 2021).
65. Chairwoman of the National Space Council, Vice President Kamala Harris announced that the United States will no longer conduct destructive tests of direct ascent ASAT missiles on 18 April 2022. Through the end of November 2022, nine other states (Australia, Canada, France, Germany, Japan, New Zealand, South Korea, Switzerland, and the UK) made similar pledges. On 7 December 2022, 155 states in the United Nations General Assembly voted in

momentum toward greater consensus and more specifics on what constitutes responsible behavior in space, which will facilitate the "naming and shaming" of parties that do not act in responsible ways. Nonetheless, these efforts also highlight just how far the space governance regime is away from governance regimes in other domains that include much more specific obligations and robust verification mechanisms, rather than voluntary guidelines and pledges.

In a most disturbing scenario, the efficacy of commercial LEO satellites in supporting Ukraine and President Putin's nuclear threats could lead the Russians to assess that the greatest military effectiveness from the use of nuclear weapons would be to detonate just one in LEO. A high-altitude nuclear detonation (HAND) would raise the peak radiation flux in parts of the Van Allen radiation belts by 3-4 orders of magnitude, cause the failure in weeks to months of *all* LEO satellites not specifically hardened against this threat, result in total financial damages probably approaching $500 billion, and present daunting response challenges since the attack would be outside of any state's sovereign territory and not directly kill anyone.[66]

Space doctrine provides a final large set of factors that shape concepts associated with space superiority in significant ways. The USSF is now responsible for developing space doctrine, and there is nothing it will do that is more important and challenging than

---

favor of a resolution calling for a halt for this type of ASAT testing while nine states voted against the resolution and nine states abstained.

66. Defense Threat Reduction Agency, "High Altitude Nuclear Detonations (HAND) against Low Earth Orbit Satellites ('HALEOS')," Advanced Systems and Concepts Office, April 2001. No satellites are known to be hardened against these nuclear effects. For further details about the threat from HAND and discussion about a potential licensing requirement for commercial LEO satellites to be hardened against residual radiation effects following a HAND, see Peter L. Hays, "United States Military Space: Into the Twenty-First Century," INSS Occasional Paper 42, (Maxwell AFB, AL: Air University Press, 2002), 101-103.

developing appropriate doctrine and culture. Simple, clear, and strongly held doctrine orients a military and is a sine qua non for creating a strong and distinct military culture. Creation of the USSF as an independent military organization should accelerate the development of space doctrine and help move it beyond what are, arguably, some of the most tenuous but often invoked analogies from maritime and air doctrine with their emphasis on controlling sea lines of communication or commanding the air. Over time, as the space doctrine development process matures, it should provide an excellent foundation for the generation-long informal processes that will incubate a space-minded culture for the USSF.

While Max Weber's and Robert Michel's iron laws of bureaucracy and oligarchy tell us that military organizations are likely to promote doctrines that build upon and expand their turf, less bureaucratic processes for doctrine development may indicate that organizations need to be expanded, reduced, or that new organizations should be created.[67] Duke University historian I.B. Holley, Jr. emphasized inherent links between doctrine and organizations in his concise definition of doctrine as "what is officially believed and taught about the best way to conduct military affairs."[68] In addition, doctrine – like strategy itself – can be thought of as theory. Good doctrine will perform the primary roles of any theory: description, explanation, and prediction. The critique that generals are focused on winning the last war has some validity – most militaries are far from ideal organizations for thinking in innovative ways about new doctrine or responding to technical and environmental changes. This is because

67. Max Weber, *Political Writings* (Peter Lassman ed., Ronald Speirs trans., 1994); Robert Michels, *Political Parties: A Sociological Study of Oligarchical Tendencies of Modern Democracy* (first published 1915, Eden Paul and Cedar Paul trans., 2001).
68. Dennis M. Drew, "Of Leaves and Trees: A New View of Doctrine," *Air University Review* (January-February 1982): 40-48; I.B. Holley, Jr., "An Enduring Challenge: The Problem of Air Force Doctrine," *The Harmon Memorial Lecture Series in Military History* (Colorado Springs: USAF Academy, 1974).

of their hierarchical structures, heightened senses of duty and tradition due to self-selection processes, and the serious consequences of making mistakes when using deadly force to achieve political objectives.

Dennis Drew's doctrine tree metaphor builds on Holley's definition and provides a useful framework for addressing two key issues related to military doctrine.[69] The first concerns the scope of doctrine and whether it should consist of fundamental principles of war applicable across different military branches, different states, and different times, or be more narrowly crafted to apply to a specific Service within a specific state at a specific time. A second set of similar issues relates to the role of the organization in formulating doctrine and the role of doctrine in organizational change. Drew posits that the different ideas within the concept of doctrine can best be thought of as parts of a tree, where the roots represent the historical lessons from which doctrine is drawn, the trunk fundamental doctrine, the branches environmental doctrine, and the leaves organizational doctrine. He defines fundamental doctrine as relatively abstract concepts such as "beliefs about the purposes of the military, the nature of war, [and] the relationship of military force to other power instruments. ..."[70] Drew also emphasizes the "timeless" nature of fundamental doctrine and indicates that it "is relatively insensitive to political philosophy or technical change."[71] Environmental doctrine "is a compilation of beliefs about the employment of military forces within a particular operating medium."[72] Moreover, environmental doctrine is "narrower in scope than fundamental doctrine because it deals with the exercise of military power in a particular medium" and is "significantly influenced by factors such as geography and

---

69. Drew, 16, 40-43.
70. Ibid., 43.
71. Ibid., 44.
72. Ibid.

technology."[73] Finally, organizational doctrine "is best defined as basic beliefs about the operation of a *particular* military organization or group of closely linked military organizations."[74] Drew indicates that organizational doctrine "is *very* narrow in scope" and that it "tends to change relatively frequently in order to remain 'current."[75] Drew's tree metaphor also illustrates the interrelationships between the different forms of doctrine: a sturdy trunk of fundamental doctrine must be developed before environmental doctrine branches or organizational doctrine leaves can grow. Similarly, leaves blow in the wind and fall off as a normal part of the life cycle of the tree, but a disease in the roots or the lower levels of the tree will stunt growth at higher levels.

To support development of a more comprehensive, well accepted, and effective doctrine for space superiority, the USSF should apply Drew's metaphor and focus sequentially on first developing doctrine rooted in universally applicable fundamental principles, then doctrine for the space environment, and finally organizational doctrine specially adapted to space. This will help the USSF avoid one set of missteps the U.S. Air Force (USAF) made in its development of space doctrine during the Cold War, when it attempted to develop space organizational doctrine without first promulgating coherent environmental doctrine for space – an approach analogous to attempting to grow leaves without a supporting branch.[76] Today, it is not clear that the range of fundamental and environmental space doctrine issues first identified during the Cold War have received enough attention, been resolved, or provide a solid foundation for understanding space superiority. There has been some progress in more recent issuances of U.S.

---

73. Ibid.
74. Ibid., 45 (emphasis in original).
75. Ibid. (emphasis in original).
76. Charles D. Friedenstein, "The Uniqueness of Space Doctrine," *Air University Review* (November-December 1985): 13-23.

military space doctrine focused on the space environment and space operations such as the October 2020 version of Joint Publication 3-14, *Space Operations*, developed by the Joint Staff,[77] and the USAF's January 2021 Air Force Doctrine Publication 3-14, *Counterspace Operations*.[78] But the United States still lacks definitive answers to several key environmental doctrine issues associated with space superiority, such as whether space is an inherently offensive domain.

Of course, because military personnel have very seldom operated in outer space and have never fought a war in space, the USSF faces very different and difficult challenges as it works to develop independent doctrine for space superiority. Sailors and pilots played an essential role in the development of maritime and air doctrine, and they could build from at least decades if not centuries of routine military operations at sea or in the air. The USSF must focus on doctrine, think creatively, and draw from all possible sources in developing doctrine to deter and win space conflicts and, despite all the logical shortcomings with this approach, look backward to see what doctrine for operations at sea and in the air may have to offer for space. Seminal theorists who developed important perspectives on military operations in these two domains include: Alfred Mahan, Julian Corbett, Giulio Douhet, Billy Mitchell, and John Warden.[79]

---

77. Joint Publication 3-14.
78. Curtis E. Lemay Center for Doctrine Development and Education, "Air Force Doctrine Publication 3-14 Counterspace Operations," (25 January 2021).
79. Alfred Thayer Mahan, *The Influence of Sea Power Upon History*, 1660–1783 (1890); Julian S. Corbett, *Some Principles of Maritime Strategy*, (first published 1911, Eric J. Grove ed., 1988); Giulio Douhet, *The Command of the Air*, (first published 1921, Richard H. Kohn and Joseph P. Harahan eds., 1983); William Mitchell, *Winged Defense: The Development and Possibilities of Modern Airpower – Economic and Military* (first published 1925, 1988); John A. Warden III, *The Air Campaign: Planning for Combat* (Pergamon-Brassey's, 1988); Jon Testuro Sumida, *Inventing Grand Strategy and Teaching Command: The Classic Works of Alfred Thayer Mahan Reconsidered* (Baltimore: Johns Hopkins University Press, 1997); Philip S. Meilinger, *The Paths of Heaven: The Evolution of Airpower Theory* (Maxwell AFB, AL: Air University Press, 1997);

Some of the key concepts that these theorists developed or applied to the air and sea domains are command of the sea, command of the air, sea lines of communication, common routes, choke points, harbor access, concentration and dispersal, and parallel attack.[80] Several of these concepts have been appropriated directly into various strands of embryonic space-power theory; others have been modified slightly, then applied. For example, Mahan's and Corbett's ideas about lines of communications, common routes, and choke points have been applied quite directly onto the space domain. Maritime and airpower concepts that have been modified to help provide starting points for thinking about space doctrine include harbor access and access to space, and command of the sea or air and space control. Unfortunately, clear and well-developed linkages between concepts like command of the sea or air and space superiority have yet to be developed and it is uncertain whether they will emerge given distinctions between the three domains and continuing evolution of space technologies.

The greatest shortfall in current space doctrine is that it lacks anything like the simple, clear, and strongly held mantra that guided early airpower advocates: *airpower is inherently offensive, manifestly strategic, and should, therefore, be organized independently*.[81] To be sure, the airpower doctrine pioneers did not always get things right, and their initial concepts were sorely tested by widespread deployment of radar, difficulties with bombing accuracy and in achieving strategic effects, and the efficacy of air defenses. As the USSF struggles to develop a simple yet powerful overarching doctrine, it must consider evolving technology developments and continue to wrestle with many fundamental

---

and David R. Metz, *The Air Campaign: John Warden and the Classical Airpower Theorists* (Maxwell AFB, AL: Air University Press, 1999).

80. John J. Klein, *Space Warfare: Strategy, Principles and Policy* (Milton Park, UK, Routledge, 2006).

81. Metz, *The Air Campaign*, provides a helpful overview of this mantra for airpower pioneers.

questions stemming from the environmental attributes of the space domain and problems with analogizing directly from sea or air doctrine. Few concepts from maritime theory translate directly into airpower theory – why should we expect either maritime or airpower theory to apply directly for the distinct domain of space?

Questions about the applicability of doctrine from other domains should not be considered a barrier to developing space doctrine, it is a reminder that doctrine development is a journey, not a destination. The purpose of this journey is to hone the minds of operators through debate and testing of nascent concepts, not to get doctrine "right." Sir Michael Howard emphasized the need for evolution and adaptation in doctrine development: "I am tempted indeed to declare dogmatically that whatever doctrine the Armed Forces are working on now, they have got it wrong. I am also tempted to declare that it does not matter that they have got it wrong. What does matter is their capacity to get it right quickly when the moment arrives. ... [It] is the task of military science in an age of peace to prevent the doctrines from being too badly wrong."[82] These sentiments are echoed by General Jay Raymond, the first Chief of Space Operations, in his foreword to the Space Capstone Publication: "This capstone doctrine is a point-of-departure toward that goal, not a final destination. Given the nascent state of space-power theory, this publication will inevitably evolve over time as it is applied, evaluated, and refined. Therefore, military space forces are encouraged to read, critique, debate, and improve upon the ideas that follow."[83]

Misuse of analogies and stifled debates clearly contributed to another set of USAF mistakes that stunted understanding of space superiority. For most of the Cold War, the USAF insisted that space

---

82. Michael Howard, "Military Science in the Age of Peace," *RUSI Journal* (March 1974).
83. John W. Raymond, *Spacepower: Doctrine for Space Forces* (Washington: Headquarters United States Space Force, June 2020), iii.

and air formed a seamless operational domain which it defined as "aerospace," a position opposed by the rest of DoD that saw distinct space and air domains.[84] Under the seamless aerospace concept, for decades the USAF tended to "force-fit" space doctrine into the mold of air doctrine and argued, inappropriately, that the three major airpower characteristics of speed, range, and flexibility, applied equally well to space-power when, in fact, speed and range mean very different things in space than in the air, and spacecraft are among the least flexible of all today's military systems. Fortunately, the USSF can now build on the much more appropriate space characteristics in the joint publication on space operations that include freedom of action; overflight; global perspective, responsiveness, multi-user capacity; and speed, reach, and persistence.[85]

Another underdeveloped construct for developing doctrine for space superiority is building from continental maritime theorists such as Raoul Castex and others, instead of continuing initial focus on open-ocean (or blue-water) operations such as those emphasized by Mahan.[86] Rebalancing toward a more limited or continental perspective on military space operations would align more closely with the current characteristics and attributes of space operations, particularly in LEO or the cosmic coastline, that are highly vulnerable to attack from Earth. Later, it may become more appropriate to rebalance thinking about space superiority back toward Mahan and other blue-water maritime theorists as space capabilities mature, conflicts over space resources intensify, and the

---

84. Peter Hays and Karl Mueller, "Going Boldly – Where? Aerospace Integration, the Space Commission, and the Air Force's Vision for Space," *Aerospace Power Journal* (Spring 2001): 34-49; Smith, "Ten Propositions," 5-7.
85. Joint Publication 3-14, I-4 and I-5.
86. John J. Klein, *Understanding Space Strategy: The Art of War in Space* (Milton Park, UK, Routledge, 2019); and Bowen, *War in Space*.

potential for large-scale, highly maneuverable space combat emerges in higher orbits and cislunar space.

Why the United States Needs Space Superiority

The short answer to why the United States needs space superiority is that every major military system and operation relies on at least some space support, if they are not completely dependent on uninterrupted, high fidelity, and assured access to terrestrial force enhancement capabilities provided by space systems. USAF Chief of Staff General Merrill McPeak described Operation Desert Storm in 1991 as "the first space war"[87] to highlight the critical importance of space support in enabling the strikingly lopsided coalition victory in that conflict. Since 1991, the importance of space support to Joint Force operations has grown exponentially so that every aspect of United States force structure, acquisition, training, doctrine, and concepts of operations is predicated on robust space support. In some cases, the United States simply could not operate without these capabilities, in many cases it could not operate in effective and efficient ways, and without space superiority it cannot assure the space support vital to these operations. Simply put, those opposing U.S. pursuit of space superiority must acknowledge that they would radically undercut the effectiveness and efficiency of the U.S. military, and they should explain any viable alternative means of maintaining U.S. ability to project power globally to assure friends and allies and dissuade, deter, and defeat adversaries. The burden of proof is on those advocating that military efficacy is no longer important to the United States in an increasingly dangerous and unstable world.

---

87. Merrill A. McPeak, "Desert Storm Reinforces Military Space Directions," *Aviation Week & Space Technology*, 8 April 1991, 42.

The force structure of the U.S. military and its concepts of operations are based on strong space support as parts of the U.S. military are transformed into smaller but more lethal units that increasingly and seamlessly leverage space information to enable their operations and substitute precision for mass. Precision engagements require precise intelligence, much of which is provided by space. As a specific example, consider that the standard air delivered precision guided munition for the past 25 years, the joint direct attack munition (JDAM), normally had a 2000-pound warhead; JDAMs are being replaced with small diameter bombs, which normally carry only a 250-pound warhead, yet can often be employed more effectively and cause less collateral damage than JDAMs. Likewise, the United States ability to project power worldwide is completely dependent on space capabilities to support and enhance these operations. The United States faces many potential "away games" where adversaries enjoy interior lines of communications in potential conflicts; U.S. asymmetric reliance on space support is particularly important in these cases and may create dangerous situations where adversaries may be advantaged by attempting to degrade all space capabilities, rather than selectively targeting specific capabilities.

Understanding how space capabilities provide critical contributions to joint all-domain command and control, time-sensitive targeting, and compressing every link in the terrestrial find, fix, track, target, engage, assess (F2T2EA) "kill chain" cycle shows that these capabilities are essential to empower the reconnaissance-strike complex at the heart of the new American way of war. The U.S. military is the world's leader in delivering precision fires and continues to improve processes to create specific effects more quickly and accurately. As it refines its doctrine for achieving space superiority, a key challenge for the USSF will be developing deep understanding and enhanced capabilities needed to execute each link in a space kill chain (or perhaps it will develop new concepts to drive space engagement cycles as capabilities evolve and become more autonomous).[88]

Finding enemy forces has been a difficult task for militaries throughout history; aircraft revolutionized reconnaissance within a theater of operations and intelligence, surveillance, and reconnaissance (ISR) collection from satellites has revolutionized it globally. Change detection enabled by sun-synchronous LEO imagery satellites often provides some of the most important contributions to ISR and enhances understanding of adversary capabilities as they develop over time. Today, pervasive global imagery provided by National Reconnaissance Office (NRO), DoD, and commercial satellites means that no military significant activities can remain hidden, and if they can be detected, they can be engaged.

Fixing involves identifying potential targets from the finding step and geolocating them. In the past, this was sometimes a difficult and time-consuming process; today the GPS system enables near-instantaneous geolocation anywhere in the world and makes a critical contribution to the kill chain. Tracking requires maintaining sensor custody of targets as they move. Persistent space imagery can make important contributions in maintaining custody of targets, but it can also be a challenge to fuse together data from various distributed space and non-space sensor networks. Comprehensive datasets of geospatial intelligence collected over decades from space and non-space sources contribute to target tracking in important

---

88. Raj Agrawal and Christopher Fernengel, "The Kill Chain in Space: Developing a Warfighting Mindset," *War on the Rocks*, 24 October 2019; Mike Benitz, "It's About Time: The Pressing Need to Evolve the Kill Chain," *War on the Rocks*, 17 May 2017; and Ridge R. Flick, "Winning the Counterland Battle by Enabling Sensor-to-shooter Automation," *Air Land Sea Space Application Center*, 1 November 2021.

ways by indicating most likely travel routes for military systems and identifying terrain obstacles and areas where they cannot operate.

Targeting is a complex link in the cycle that is enabled by several space capabilities. Most importantly, military satellite communications, and perhaps commercial satellite communications, are usually needed to connect the entire kill chain together and allow commanders to control the operation. Commanders determine target priorities and sequencing, decide what capabilities will be used for engagement, and direct the forces that employ these capabilities. Depending on the time available and sensitivities regarding certain targets, commanders may use decision making tools that provide extremely detailed modeling of weapons effects based on the desired mean point of impact, predicted damage to target, and potential collateral damage.

Engagement is the use of capabilities against the target, another link in the kill chain cycle in which space capabilities provide essential contributions. The GPS system is again critical in this part of the kill chain, as most munitions use this capability to guide their employment for at least some portion of the engagement. Assessment, the final link in the kill chain is also dependent on space capabilities; this link is sometimes known as battle damage assessment (BDA). Many of the same capabilities used initially to find targets are used to assess whether the engagement achieved the desired effects. As engagements become increasingly precise and may employ non-kinetic means, it can be difficult to assess whether subtle desired effects were achieved.

Pathways to Achieve Space Superiority

When assessing potential pathways toward achieving space superiority, the USSF should consider an important construct that relates to doctrine, the need for space weapons, and organizational structures. This framework is derived from the four-part typology first developed by USAF Lieutenant Colonel David E. Lupton in 1983 and expanded for his 1988 book, *On Space Warfare*.[89] Lupton's *sanctuary* school of thought posits that the most useful military applications of space are for systems that enhance strategic stability and facilitate strategic arms control. Satellites contribute to these critical functions by monitoring the strategic forces of potential enemies, reducing the likelihood of surprise attack, and providing national technical means of verification (NTMV) for arms control agreements. Missile warning satellites, such as the Space-Based Infrared System, strengthen strategic stability by providing worldwide surveillance of ballistic missile launches that enhances the survivability of, and control over, retaliatory strategic forces. Other military space systems – particularly nuclear command, control, and communications supported by Advanced Extremely High Frequency satellites – are also critical for strategic stability. The sanctuary school fits very closely with the mutual assured destruction (MAD) paradigm for strategic nuclear deterrence, putatively a foundational concept during much of the Cold War. Because of the critical importance of the stabilizing functions performed by spacecraft, proponents of the sanctuary school believe that space must be kept free of weapons, and they are especially concerned with prohibiting ASAT weapons that threaten spacecraft performing these vital functions. Critics of this school charge that it misunderstands how important space capabilities have become in terrestrial military operations, minimizes the implications of many space systems that are dual use, and lacks appreciation for the growing number of dedicated ASAT systems and residual ASAT

---

89. David E. Lupton, "Space Doctrines," *Strategic Review* (1983): 36-47; Lupton, *On Space Warfare: A Space Power Doctrine* (Maxwell AFB, AL: Air University Press, 1988).

capabilities.[90] Counterintuitively, by removing a threat to such systems, an effective ASAT ban might even foster an environment conducive to the development of very threatening and destabilizing space systems, including potential space-to-Earth strike systems.[91]

The *survivability* school is focused on improving the resilience of space systems. It is clearly related to the sanctuary school because it also sees the ability of spacecraft to enhance stability as their most important function. However, it evolves away from the sanctuary school by arguing that technological developments indicate that space can no longer be maintained as a sanctuary and emphasizing that space systems deployed to promote stability also have significant ability to enhance the military effectiveness of terrestrial forces. This school derives its name from the idea that space systems are inherently less reliable, supportable, and survivable than are terrestrial forces, and must therefore specifically be designed and deployed in ways to make them more robust and resilient. In September 2015 the Office of the Secretary of Defense issued a white paper explaining ways to improve space mission assurance including defensive operations, reconstitution, and resilience; resilience approaches were further divided into disaggregation, diversification, deception, protection, proliferation, and distribution.[92]

It is helpful to build from Lupton and consider current developments to draw clearer distinctions between the four schools of thought.

---

90. Of course, discussions about dominating space, describing space as a warfighting domain, and creation of the USSF are clear repudiations of the sanctuary school.
91. Carter describes this as the paradox of ASAT arms control; see Ashton B. Carter, "Satellites and Anti-Satellites: The Limits of the Possible," *International Security* (Spring 1986): 68.
92. Office of the Assistant Secretary of Defense for Homeland Defense & Global Security, "Space Domain Mission Assurance: A Resilience Taxonomy," (September 2015). These methods of enhancing resilience are also discussed in Joint Publication 3-14.

Under this approach, the survivability school of thought can focus on just non-offensive or passive defensive measures, positing that they can be sufficient to enhance survivability and that active defense measures to protect friendly space capabilities from attack are not needed. This is a sharper distinction than in current doctrine under which both active and passive defense actions are included in measures referred to as defensive space control in joint doctrine, as defensive counterspace operations in USAF doctrine, and as defensive operations in USSF doctrine. Drawing these fine distinctions in practice is difficult and fraught with many contentious opinions as indicated by the decades-long debates about issues like anticipatory self-defense. Critics of the survivability school question whether purely defensive approaches can be effective and efficient, are uncertain whether space systems are inherently more vulnerable than other types of military systems, and some oppose the restrained approach to military space advocated by this school.

The third school of thought identified by Lupton is the *control* school, which holds that space should be thought of like other military theaters of operation where the primary military objective is to gain control over the domain.[93] "Control" implies an ability to maintain one's freedom of action while also having the ability to deny freedom of action to adversaries. In the expanse of space, it is unlikely that one can exert control over large areas for extended times; control is more likely to be exercised over limited areas for specific times. Space strategists also draw distinctions between temporary and reversible means to gain control versus means of gaining control that cause permanent effects. More specifically, they envision a scale of actions moving from temporary and reversible effects toward permanent effects that includes deception, disruption, denial, degradation, and destruction.[94] The space control school

93. Lupton, *On Space Warfare*, 60-70.
94. Air Force Doctrine Publication 3-14, 1.

posits that both offensive and defensive operations are important and likely to be conducted in space. The USAF defines offensive counterspace operations as those "undertaken to negate an adversary's use of space capabilities, reducing the effectiveness of adversary forces in all domains."[95] While space control can be considered independently, this concept is often linked with helping to accomplish military missions from space such as terrestrial force enhancement and force application; it could also be associated with non-military functions such as exploration and commercial exploitation of space. Critics of the space control school charge that it encourages an expensive and unnecessary arms race in space that they believe would not enhance security on Earth.[96]

Lupton's final school on the military utility of space holds that space clearly has the potential to be the decisive theater of combat operations. Reasoning by historical analogy, the *high ground* school posits that just as holding the high ground is often a key factor in a land battle or as airpower often prevails over land and sea forces, in the future, space forces will dominate terrestrial forces. Lupton, along with many other analysts in the 1980s, linked the high ground school directly with the strategic defense initiative (SDI) and the concept of space-based ballistic missile defense (BMD).[97] Conceptually, however, the high ground school is broader than legacy strategic debates and envisions force application missions from space for more than just BMD. As the widespread debate over SDI indicated, many oppose these high ground concepts for several reasons including: their undermining effect on MAD as a basis for strategic stability, the alleged extreme expense and technological

---

95. Ibid.
96. See, for example, Nancy Gallagher and John David Steinbruner, "Reconsidering the Rules for Space Security." (Cambridge: American Academy of Arts & Sciences, 2008).
97. Lupton, *On Space Warfare*, 52–59.

barriers involved, and the likelihood of stimulating a wide-open arms race in space.

Table 1 expands on Lupton's concepts and provides brief descriptions of space system characteristics and potential employment strategies, likely combat missions for space forces operating in accordance with each of the schools, and the types of military organizations for space operations and advocacy usually desired by the proponents of each school. Several factors including traditional military preferences for offensive doctrine in all domains, make it likely the creation of the USSF will accelerate the trend toward the control and high ground schools that has emerged over the past 30 or more years.

Table 6.1 Attributes of Military Space Doctrines

	Primary Value and Functions of Military Space Forces	Space System Characteristics and Employment Strategies	Conflict Missions of Space Forces	Appropriate Military Organization for Operations and Advocacy
Sanctuary	• Enhance Strategic Stability • Facilitate Arms Control	• Limited Numbers • Fragile Systems • Vulnerable Orbits • Optimized for NTMV Mission	• Limited	National Reconnaissance Office
Survivability	Above functions plus: • Force Enhancement	• Terrestrial Backups • Commercial and International Augmentation • Autonomous Control • Attack Warning Sensors • Less Vulnerable Orbits • Hardening • Redundancy • Crosslinks • Maneuver • Space Mission Assurance • Defensive Operations	• Force Enhancement • Degrade Gracefully	Major Command or Combatant Command

Control	• Control Space
	• Significant Force Enhancement
		• Resilience:
		  • Disaggregation
		  • Protection
		  • Distribution
		• Proliferation
		• Diversification
		• Deception
		• Reconstitution
		• On-Orbit Spares
			• Control Space
			• Significant Force Enhancement
			• Surveillance, Offensive, and Defensive Counterspace
				Combatant Command, Space Corps, or Space Force

High Ground	Above functions plus:
	• Decisive Impact on Terrestrial Conflict
	• Ballistic Missile Defense (BMD)
		• 5Ds:
		  • Deception
		  • Disruption
		  • Denial
		  • Degradation
		  • Destruction
		• Bodyguards and Convoys
			Above functions plus:
			• Decisive Space-to-Space and Space-to-Earth Force Application BMD
				Space Corps or Space Force

As indicated in the space system characteristics and employment strategies column in Table 1, a wide range of options that would likely pose increasing threats to others are available to support operations under the survivability, control, and high ground schools. Returning to the analysis from Jervis and Townsend, a key factor is whether adversaries can distinguish between offensive and defensive capabilities and whether they perceive that the offense or the defense has the advantage in space. To avoid doubly dangerous orbital security dilemmas, space architectures and employment strategies should emphasize non-offensive or passive defenses whenever feasible. Space mission assurance should be strengthened by enhancing defensive operations, resilience, and reconstitution capabilities. There are many attack vectors in space, but offensive space control capabilities provide only limited utility for strengthening space mission assurance in most scenarios. Offensive space control is needed primarily to negate enemy space capabilities targeting U.S. and allied forces, rather than as a means of active defense. Refining, clarifying, and delineating the role of offensive space control capabilities in achieving space superiority is perhaps the single most important near-term challenge for the USSF's doctrine development efforts.

If the U.S.'s space architecture and employment strategies are perceived in these non-offensive ways, war in space might be less likely. Of course, the long history of misperceptions and blundering into war does not inspire confidence that passive defense approaches alone can be sufficient to avoid war. It is therefore prudent for the United States to posture itself to win any race to deploy robust offensive space control and high ground space capabilities. Fortunately, today it is not necessary for the United States to initiate such a deployment race to maintain space superiority because there are better and less threatening options focused on accelerating acquisition and employment of far more resilient space architectures.

The best approach for pursuing and maintaining U.S. space superiority leverages several current trends including development of much more resilient space architectures, improved means of denying benefits from attacks in space, and mitigating pressures to perceive space as doubly dangerous. The United States should accelerate comprehensive and multidimensional efforts to make all space architectures more resilient. Key measures include greatly proliferating satellite numbers with mixed fleets of commercial, international, and U.S. government systems; diversifying and distributing mission functions to as many nodes in as many domains as feasible; hardening against residual radiation following a HAND; and employing enhanced deception techniques and gray zone activities. Better reconstitution capabilities with waiting satellites optimized for warfighting that fuse signals intelligence with radar imagery to provide real-time engagement solutions for proliferated moving targets would deter initial space attacks. Protective measures must address both kinetic and directed energy threats more effectively and efficiently; the Nitze criteria that defenses should be cost-effective on the margin is a useful concept for assessing the efficiency of protective measures. Building the space analogue of the multidimensional defense-in-depth strategies familiar for air and sea defense will enhance the USSF's ability to deliver capabilities better able to fight through attacks and absorb attrition while still delivering critical support to terrestrial forces.

Enhanced space domain awareness is essential to advance U.S. space superiority, strengthen deterrence of attacks in space, and improve integrated deterrence. Work to transition space safety and space traffic management from DoD to the Department of Commerce (DoC) should be accelerated and will better enable DoD to focus on intelligence support for operations in the contested space domain. Both the DoC and DoD efforts should move away from legacy processes, use state-of-the-art astronomy techniques, ingest and vet data from many more sources, and employ big data analytics more effectively.

Finally, it is incumbent on the USSF to debate, test, refine, and advance all concepts associated with space superiority. Enhanced focus on this core competency will help ensure U.S. space superiority, improve integrated deterrence, and manifest a brighter space future for humanity.

Bluewater and Brownwater Space Strategies and Their Budgetary Profiles

Peter Garretson

This essay examines how different ideas about military space-power have force structure and resource implications. It reviews the underlying assumptions of the two leading 'schools' of space-power theory and how those assumptions generate divergent preferences and allocation of resources across rhetoric, planning, force design, and budget. It also explores whether these different assumptions might lead to fundamentally different levels of societal resourcing as well, in essence not just the different ways the Space Force pie might be cut, but differences in the size of pie as well.

Whither Space Force Acquisitions and Force Design?

Since December of 2019, America has a Space Force.[98] There is tremendous focus on reforming acquisition so the Space Force can move fast.[99] But what systems should it be acquiring for what anticipated future? It turns out that there are fundamental strategic

98. Jim Garamone, "Trump Signs Law Establishing U.S. Space Force," U.S. Department of Defense, December 20, 2019, https://www.defense.gov/News/News-Stories/Article/Article/2046035/trump-signs-law-establishing-us-space-force/.
99. Theresa Hitchens, "For Space Force, It's Acquisition, Acquisition, Acquisition: 2022 Preview," *Breaking Defense* (blog), December 29, 2021, https://breakingdefense.com/2021/12/for-space-force-its-acquisition-acquisition-acquisition-2022-preview/.

choices which lead to different futures and different postures. What anticipation of different theaters of strategic competition have a tiny impact in the near-term, they become increasingly divergent over the longer term.

The Ideological Split: Looking Down vs. Looking out

In the NPEC publication, *What a War in Space Might Look Like Circa 2030-2040,* [100] two very different archetypes of conflict in space were explored as expounded by two space-power theory communities: a conservative, downward-looking "brownwater" vision of space conflict, and an outward-looking, "bluewater" vision of how space conflict might take shape.

Building on this earlier paper, here we explore the budgetary, investment, and force structure implications of these two schools of space-power theory. It is proposed that accepting the assumptions of one theory versus the other leads to different attitudes toward emergent problems, and different levels of resourcing. Even more fundamentally, differing assumptions lead to different 'contracts' with the American people, potentially resulting in very different levels of resourcing for military space forces.

In essence, we can differentiate the views of the two schools of space-power theory as follows:

The conservative "brownwater" school of space-power conceptualizes the value of military space-power primarily in terms of "looking down" on planet Earth to provide critical services to the terrestrial joint force, such as overhead observation, missile

---

100. Peter Garretson, "What War in Space Might Look Like Circa 2030-2040?," NPEC, August 28, 2020, https://npolicy.org/article_file/What_War_In_Space_Might_Look_Like_In_The_Next_One_To_Two_Decades.pdf.

warning, over-the-horizon communications, command and control, and precision navigation for targeting. An analogy would be the spotting utility of early aircraft to artillery and army movement.

In contrast, the expansive “bluewater” school of space-power conceptualizes the value of military space-power primarily in terms of “looking out” from planet Earth to facilitate and protect access to new and vast energy and material resources of the inner solar system, beginning with the Moon and extending to the asteroid belt. An analogy would be the role of navies in protecting the investments of the contending European powers in establishing economic footholds in the new worlds.

Different Assumptions

Brownwater assumptions include that space is key to terrestrial warfighting, and that expansive operations will be marginal and largely the responsibility of other actors, chiefly NASA or commercial. Generally, brownwater theorists do not anticipate or find credible step changes that meaningfully alter the current problems and emphasis of military space-power. Alternately they might agree with the bluewater view but disagree about the degree of urgency of the need to invest in such capabilities. They generally assume space information services will become ever more important to conventional war operations, and that counter-space operations become increasingly important to secure or deny those information services.

For brownwater theorists, the USSF contract with America is to support the Joint Force through the missions of military satellite communications (MILSATCOM), precision navigation and timing (PNT), missile warning (MW), weather observation and forecasting, space domain awareness (SDA), space control (SC), and a hoped expansion into tactical intelligence surveillance and reconnaissance (ISR), chiefly the overhead real-time surveillance of mobile systems (tanks, ships, aircraft, hypersonic and cruise missiles). In return, the USSF expects a commensurate share of the defense budget, assumed to be flat except for an annual increase each year to cope with inflation.[101]

Bluewater assumptions, in contrast, assume a step change and exponential growth in national space activities, resulting in spacefaring nations having, and competing for, important economic interests in space -- and a corresponding need to protect them. They assume that space becomes an important, perhaps even central theater for great power competition. They find credible the announced plans of states and industry to create a step-change in access (for example, SpaceX's Starship), to establish multiple permanent facilities on the Moon (for example, the U.S.' Artemis program and the PRC-Russian International Lunar Research Station). They take seriously that new industries will emerge from these activities, including extractive industries, and in-space manufacturing.

101. I used an 2.5% which now looks optimistic, given the new long-term inflation is likely to be 3-5%.

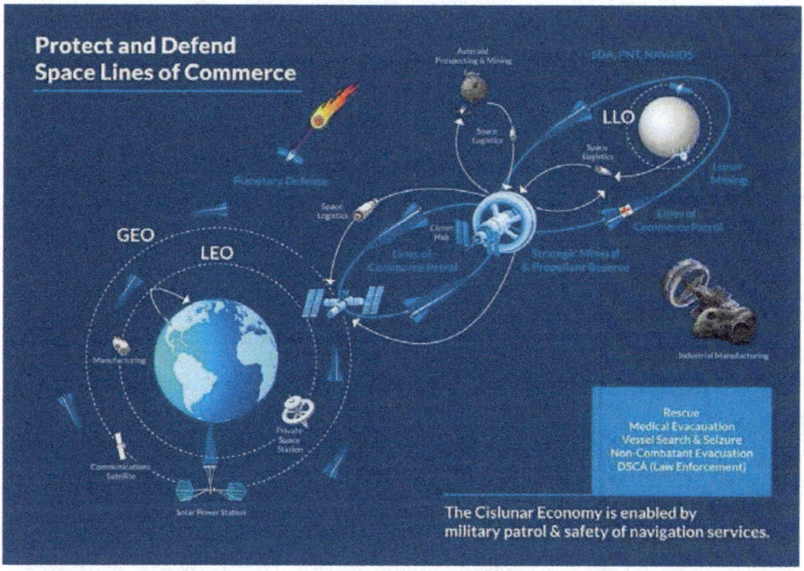

Protect and Defend Commerce (Source: SSIB2021)[102]

For bluewater theorists, the USSF contract with America is more expansive, and includes the duties to protect and defend commerce, even to proactively facilitate commerce through targeted R&D and military buying power. Its additional missions, derivative of the protection and defense of commerce, include robust Cislunar SDA, Space Control, rescue, and defense against asteroid and comet threats, and a strong commitment to the "Space Security" mission, which is a peacetime presence and assurance mission that enables safety and security of commercial, interagency, and international partners. Bluewater theorists expect military operations to expand beyond Geostationary orbit (XGEO) in tandem and in proportion

102. J. Olson et al., "State of the Space Industrial Base 2021: Infrastructure & Services for Economic Growth & National Security," ed. Peter A. Garretson (DOD, AFRL, USSF, November 2021), https://assets.ctfassets.net/3nanhbfkr0pc/43TeQTAmdYrym5DTDrhjd3/1218bd749befdde511ac2c900db3a43b/Space_Industrial_Base_Workshop_2021_Summary_Report_-_Final_15_Nov_2021.pdf.

with the civil and commercial space activities. In return, a bluewater USSF would expect to see its funding scale proportionally with the expansion of the space economy, at the same ratio of military spending to national GDP (3.2%). Over time, assuming the space economy continues compound growth, this results in a substantially larger budget.

Whither Our Force Mix?

In reality, a USSF budget will be a mix of both archetypes, seeking a balance between investments that protect national interests on Earth, and investments that protect national interests in space. The above descriptions are caricatures. Both individuals and national security space organizations are likely to lie on the spectrum in between these two extremes, agreeing in principle as to the long-term direction but disagreeing in practice as to how much and when to invest in each sphere. However, expectations matter toward the posture taken, and the resources devoted to proactively developing capabilities for each to the exclusion of the other.

Ideas Have Implications

How might this express itself? We can expect differences with regard to the posture taken toward Space Security Missions, toward the Moon, Rhetoric and Cultural References by senior USSF leadership, planning efforts, and budget allocations for operations beyond GEO (XGEO), and budget posture with respect to Congress.

Postural Differences

	Brown Focus (Flat)	Blue Focus (Flat)	Blue Focus (Pegged)
Attitude toward Space Security Missions	Averse, Hassle, Reticent	Assertive, Proactive	Assertive, Proactive
Attitude toward Moon	Reticent	Proactive	Proactive
Rhetoric & Cultural References	1%	50%	50%
Planning Effort	1%	10%	10%
XGEO Budget allocation (initial)	<1%	2%	2%

To the extent that "brownwater" ideology prevails, we would expect the USSF leadership to be largely averse to demand signals for increased Space Security mission requirements, reticent to take them on, seeing them as a distraction, hassle, and unfunded mandate that detracts from their core mission area of support to the Joint Force. The attitude toward the Moon and Cislunar development would be similarly reticent, seeing the Moon principally as NASA's purview, and only interested in Cislunar to the extent it provides an attack vector to Earth-servicing military platforms. As a result, they would only expand XGEO in response to threats. The near-term consequences are that they would de-emphasize Space Security, the Cislunar economy, and the long-term economic value proposition of Space, with senior leader rhetoric and cultural references alluding to such concerns <1% of their talking points, and usually only when asked. Moreover, the amount of planning effort devoted to XGEO and new Space Security missions compared to the total planning effort would likewise be about 1%. Similarly actual near-term fiscal investment for XGEO would be <=1% of the Space Force budget. The basic assumption regarding the long-term budget outlook is a fight with the other services over a fixed pie that is largely flat over time.

To the extent that "bluewater" ideology is influential, we would see USSF leadership taking an assertive and proactive stance with regard to Space Security missions, seeing these as a core peacetime presence mission. They might proactively attempt to codify such missions via legislative proposals, via executive roles & missions guidance, and via OSD and Joint Staff planning guidance. USSF leadership would take a proactive stance with regard to the Moon and Cislunar development, considering it a central theater of significant U.S. economic interest which must be developed, secured via presence at key terrain, proactive development of a dual-use Cislunar logistical system. As a result, they would champion early presence in tandem with industry and civil efforts, and would champion investment in technologies which enable both commerce at scale and military logistics, as well as develop formal requirements. The near-term consequences are that the USSF leadership would be aggressive about building the long-term cultural identity as a spacefaring service and building alliances with spacefaring advocates and companies, and so messages about Space Security, the Cislunar economy and the long-term value proposition of the Space Force might account for 50% of senior leader talking points, especially in posture statements. They would devote significantly more time to planning for the future, allocating on the order of 10% of planning activity to building new concepts of operations (CONOPS), requirements, and be proactive about ensuring they are tasked by higher level documents (such as the Guidance for the Employment of the Force (GEF), and Joint Strategic Capabilities Plan (JSCP). As a result, near-term budget allocation for XGEO would begin at 2% of the USSF budget, and plan to scale with commercial/civil needs. The basic assumption regarding the budget is that the U.S. space economy will be continually growing, and that to the extent the USSF is providing a valuable service, its budget will scale accordingly.

Potential Economic Implications of Ideas for Total Resourcing

Ideas may matter even more starkly when it comes to total resourcing. While a brownwater Space Force, as a "must-pay bill" supporting the Joint Force, can only lay claim to a small portion of the total defense budget: the negotiated share required to support the terrestrial warfighter, adjusted each year for inflation (say 2.5%). Only those few defense contractors who are rentiers on the budget are likely to militate for a larger budget share and must compete for shares in a fixed pie.

However, a Blue Water Space Force could "hitch its wagon to a star"— signing on to protecting the rapidly growing space economy allowing it to legitimately make a claim to a much larger budget. Because a commercial space industry services a far larger base of consumers, it has a much larger tax-base, and has access to a broader coalition for congressional influence. Moreover, both their resources and influence are steadily growing. To the extent that bluewater missions directly support their security and expansion, they are likely to support and advocate for, and therefore secure, a larger USSF budget. How might this play out?

In 2022, the Space Foundation pegged the global space economy at $469B. [103] The BEA estimates that the U.S. portion of the space economy was $194.4 billion. But the space economy is growing. It grew 10% this past year, and is expected to grow at 8% from 2022

---

103. Rich Cooper and Dottie O'Rourke, "Space Foundation Releases The Space Report 2022 Q2 Showing Growth of Global Space Economy," Businesswire, July 27, 2022,
https://www.businesswire.com/news/home/20220727005344/en/Space-Foundation-Releases-The-Space-Report-2022-Q2-Showing-Growth-of-Global-Space-Economy.

to 2027. [104] The Space Foundation expects it to reach beyond $634 billion by 2026 [105] and to exceed $1 trillion by 2040. [106]

As evidenced by The Future of Space 2060 and Implications for US Strategy,[107] at least parts of the Space Force are taking an even longer view. George Pullen has developed a model that projects the total size of the space economy, U.S. share of the space economy, and the expected relative size of the in-space economy.

---

104. Mordor Intelligence, "Spacecraft Market | 2022 - 27 | Industry Share, Size, Growth - Mordor Intelligence," Mordor Intelligence, 2022, https://www.mordorintelligence.com/industry-reports/spacecraft-market.
105. Carlie Porterfield, "Space Industry Grew To Record $469 Billion Last Year, Report Finds," Forbes, July 27, 2022, https://www.forbes.com/sites/carlieporterfield/2022/07/27/space-industry-grew-to-record-469-billion-last-year-report-finds/.
106. Citi GPS, "SPACE: The Dawn of a New Age," May 9, 2022, https://icg.citi.com/icghome/what-we-think/citigps/insights/space_20220509.
107. Joel B Mozer et al., "The Future of Space 2060 and Implications for US Strategy: Report on the Space Futures Workshop" (Air Force Space Command Peterson AFB United States, September 5, 2019), https://aerospace.csis.org/wp-content/uploads/2019/09/Future-of-Space-2060-v2-5-Sep.pdf.

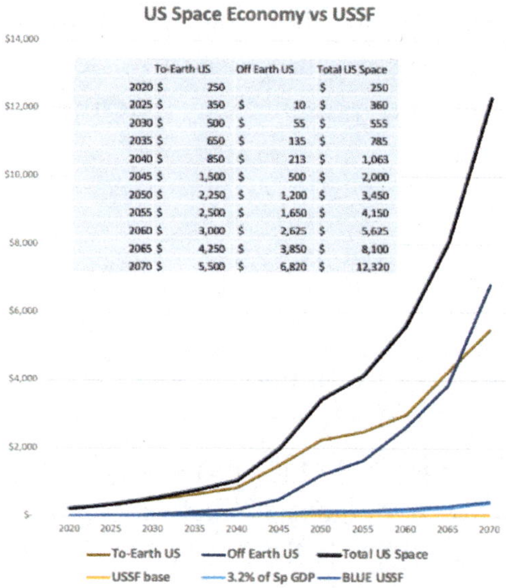

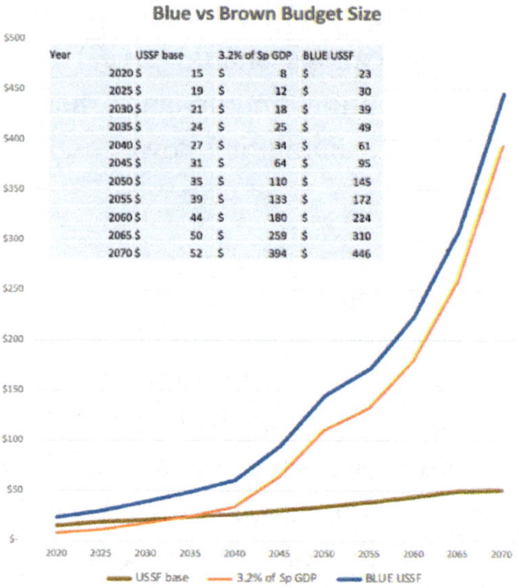

US Space Economy vs USSF budgets (left). / Blue vs Brown budget Size (right) Source: Pullen[108]

If we assume this robust growth of the space economy, and assume the US would spend proportionally an equal amount to protect the space economy as it does to protect other economically productive activity, what then is the consequence for the Space Force budget?

As a starting point, the U.S. spends 3.2% of its GDP on defense. As seen in the above figure, the brownwater or base budget of the USSF increases slowly with inflation from ~$15B in 2020 to $52B in 2070.

In stark contrast, a bluewater USSF, pegged to 3.2% of U.S. Space GDP, grows rapidly to eclipse the warfighting budget in 2035, *achieving a total budget nearly 9 times the size of the base budget in 2070 of $446B*, with about $200B of that devoted to securing the space-to-space / Cislunar economy. While any future economic

108. George Pullen, "Projections of the Future Space Economy to 2070" (Personal work done for author for this paper, 2022).

projection or scenario is speculative, the vastly divergent resourcing implications should give USSF strategists pause: a narrow self-conception might and rhetoric in the present may sacrifice significant buying power in the future.

Conclusion

Ideas have resource implications. A prevailing looking-down ideology is likely to provide scant resources for the Moon or a broader in-space economy, where a looking out ideology is likely to accelerate those efforts. A looking down ideology is likely stiff arm new missions whereas a bluewater ideology is likely to be jealous of potential new missions and aggressively seek them. A looking down ideology is likely to grow only in relation to the overall DoD budget, whereas a looking out ideology is likely to add new dollars to the DoD budget in tandem with the tax base of a broader space economy. Over time, this might result in a significant divergence, where a looking-out Space Force might be many times the size of a looking down Space Force.

The actual investment profile is likely to be a mix of both bluewater and brownwater considerations. Two different theories of the value of space-power provide different prescriptions for resource allocation, and result in two potential failure modes. The failure mode of one (bluewater) is the proactive investment in a far-term future that may not materialize and reduces the budget available for near-term warfighting capabilities. The failure mode of the other (brownwater) is a failure to invest in shaping an exponentially expanding area of national interest and being flat-footed and late-to need. The wisdom of an informed public may help shape this balance, and this paper has suggested ways in which the attentive public can recognize and evaluate the relative influence of bluewater and brownwater ideas on Space Force Resource allocation.

Appendix: Is the Moon a Strategic Destination?

At Issue: Whether the Moon is a Strategic Destination

A fundamental wedge issue between bluewater and brownwater thinking is how they conceptualize the Moon: Do they either 1) evaluate the Moon itself as a theater for strategic competition requiring military capabilities to protect U.S. economic interests,[109] or 2) evaluate the Moon's potential economic value is so remote and peripheral, that it can be safely ignored to concentrate on terrestrial concerns. [110]

It is perhaps natural for the layperson to dismiss the Moon. The received wisdom has been that the Moon, while symbolic in the 1960s, turned out to be just a dry, dusty rock, to which the United States did not even bother returning after 1972. [111] However, it is at least worth considering that the future will not look like the past, and that the Moon might be strategically important – by which we mean that it could affect the wealth of nations, the ability to control or constrict such wealth, and thus who may be the leader or leading states in the macro-political international system.

Those who don't closely follow all aspects of space – civil, commercial, and military – are likely unaware, and may be surprised

---

109. David Buehler et al., "Posturing Space Forces for Operations Beyond GEO," *The Space Force Journal* 1, no. 1 (January 31, 2021), https://spaceforcejournal.org/posturing-space-forces-for-operations-beyond-geo/.
110. Theresa Hitchens, "Kendall's 'message' to Space Force: Support Missions Are Central Role," *Breaking Defense* (blog), April 6, 2022, https://breakingdefense.sites.breakingmedia.com/2022/04/kendalls-message-to-space-force-support-missions-are-central-role/.
111. RMG, "Why Did We Stop Going to the Moon?," Royal Museums Greenwich, February 27, 2021, https://www.rmg.co.uk/stories/topics/why-did-we-stop-going-moon.

to learn just how much near-term activity will be taking place on, or near the Moon. In fact, over 100 missions (mostly robotic) are planned to the Moon in the next decade.[112] More than half of which are commercial. China, Russia, Japan, India, Israel, Germany, Canada, South Korea, the United Kingdom, the United States and the European Space Agency (ESA) all have planned near-term missions. There is substantial evidence that nations are already looking at the Moon in terms of economic resources for great power competition rather than for prestige.[113] The most ambitious state programs belong to China and the United States.

China's efforts are not directed at prestige, but at creating an industrial and manufacturing base on the Moon[114] which they hope will lead toward a Moon-Earth Economic Zone that will annually produce $10 Trillion by 2050[115] [Note, that is 10% of today's global GDP]. The PRC has taken a number of concrete steps toward this goal. They have already sent multiple orbiters and landers, as well as a relay station at Earth-Moon-Lagrange 2 to enable them to explore the far-side and shadowed craters on the poles. They have succeeded in Lunar sample return.[116] Their near-future plans are to conduct a Lunar sample return on the South Pole, and demonstrate 3D printing on the Moon. The PRC has announced plans to establish

---

112. Olson et al., "SSIB 2021."
113. Namrata Goswami and Peter A. Garretson, *Scramble for the Skies: The Great Power Competition to Control the Resources of Outer Space* (Lexington Books, 2020), https://rowman.com/ISBN/9781498583114/Scramble-for-the-Skies-The-Great-Power-Competition-to-Control-the-Resources-of-Outer-Space.
114. Xinhua, "Exploiting Earth-Moon Space: China's Ambition after Space Station - China - Chinadaily.Com.Cn," China Daily, March 8, 2016, https://www.chinadaily.com.cn/china/2016-03/08/content_23775949.htm.
115. Cao, Siqi, "China Mulls $10 Trillion Earth-Moon Economic Zone - Global Times," Global Times, November 1, 2019, https://www.globaltimes.cn/content/1168698.shtml.
116. Stephen Clark, "Chinese Mission Returned Nearly 4 Pounds of Lunar Samples – Spaceflight Now," January 1, 2021, https://spaceflightnow.com/2021/01/01/chinese-mission-returned-nearly-4-pounds-of-lunar-samples/.

an International Lunar Research Station (ILRS) in concert with Russia, with the construction phase beginning in 2030, and invited all nations to participate,[117] even releasing an international partnership guide.[118] Their long-term plan is to industrialize the Moon to a point where they can build solar power satellites[119] – enormous orbital power stations they see as their Apollo program equivalent which will make them a global leader in energy and Aerospace.[120]

America also has Lunar plans. Largely spurred by the PRC's success in landing on the far side of the Moon, the White House has directed NASA to return humans to the Moon by 2024 and to establish a permanent presence by 2028. While NASA will miss both directed dates by several years, it has successfully launched its first uncrewed circum-Lunar mission this year,[121] and is expected to return humans to the Moon by 2026, and establish a permanent presence in the 2030's. Augmenting the main Artemis program is the Commercial

---

117. Andrew Jones, "China, Russia Open Moon Base Project to International Partners, Early Details Emerge," *SpaceNews* (blog), April 26, 2021, https://spacenews.com/china-russia-open-moon-base-project-to-international-partners-early-details-emerge/.
118. CNSA, "International Lunar Research Station (ILRS) Guide for Partnership," China National Space Agency (CNSA), June 16, 2021, http://www.cnsa.gov.cn/english/n6465652/n6465653/c6812150/content.html.
119. Xinhua, "Exploiting Earth-Moon Space: China's Ambition after Space Station - China - Chinadaily.Com.Cn."
120. Economic Times, "China Plans to Build Huge Space Solar Power Station," *The Economic Times*, March 30, 2015, https://economictimes.indiatimes.com/news/international/business/china-plans-to-build-huge-space-solar-power-station/articleshow/46742968.cms?from=mdr.
121. George Dvorsky, "7 Things We Learned From NASA's Wildly Successful Artemis 1 Mission," Gizmodo, December 14, 2022, https://gizmodo.com/nasa-artemis-orion-sls-moon-mission-1849892615; Mike Wall, "NASA over the Moon with Success of Artemis 1 Orion Mission | Space," Space.com, December 11, 2022, https://www.space.com/nasa-celebrates-artemis-1-orion-success; Rachel Kraft, "Analysis Confirms Successful Artemis I Moon Mission, Reviews Continue," Text, NASA, March 7, 2023, http://www.nasa.gov/feature/analysis-confirms-successful-artemis-i-moon-mission-reviews-continue.

Lunar Payload Service (CLPS) program[122] to deliver small and medium payloads, and the Lunar Surface Innovation Initiative (LSII) which is exploring industrial-relevant technologies.[123] More importantly, the United States selected the South Pole because of its resource value,[124] and has issued a National Cislunar Strategy which seeks to grow infrastructure and economy, and demonstrate scalable extraction and manufacturing processes using in-situ materials.[125]

The Artemis precursor missions use small commercial landers to begin resource mapping. Its selected Human Landing System (HLS),[126] SpaceX Starship will have the capability to emplace 100MT on the Moon, which exceeds earlier estimates of the necessary mass for a self-sustaining and growing industrial base. Starship will massively drop the cost of access to space while simultaneously increasing the payload.[127] Several U.S. companies have plans for extractive industry on the Moon, and one is already manifested on an HLS which will enable significant levels of production.

---

122. Gary Daines, "Commercial Lunar Payload Services," Text, NASA, March 14, 2019, http://www.nasa.gov/content/commercial-lunar-payload-services.
123. Loura Hall, "Lunar Surface Innovation Initiative," Text, NASA, October 22, 2019, http://www.nasa.gov/directorates/spacetech/Lunar_Surface_Innovation_Initiative.
124. SpaceRef Editor, "Recommendations Approved by the National Space Council to President Trump," SpaceRef, March 26, 2019, https://spaceref.com/status-report/recommendations-approved-by-the-national-space-council-to-president-trump-2/.
125. White House and NSTC, "National Cislunar S&T Strategy" (The White House, 2022), https://www.whitehouse.gov/wp-content/uploads/2022/11/11-2022-NSTC-National-Cislunar-ST-Strategy.pdf.
126. Jeff Foust, "NASA Selects SpaceX to Develop Crewed Lunar Lander - SpaceNews," SpaceNews, April 16, 2021, https://spacenews.com/nasa-selects-spacex-to-develop-crewed-lunar-lander/.
127. Peter Garretson and Samuel Havard, "The Starship Singularity" (AFPC, February 22, 2023), https://www.afpc.org/uploads/documents/Starship_(SPR)_-_2.22.23.pdf.

At the same time NASA has two planned missions, OSAM-1 and OSAM-2 to develop basic competencies for in-space servicing, assembly and manufacturing, and the Biden administration has released an In-Space Servicing Assembly and Manufacturing (ISAM) strategy[128] to "put factories in space."[129] Very recently, NASA has released its Moon-to-Mars (M2M) objectives, which provide a number of Lunar Industrialization objectives,[130] including the demonstration of "advanced manufacturing and autonomous construction capabilities" and "industrial scale ISRU capabilities," in support of "continuous human lunar presence and a robust lunar economy" as well as "lunar power generation and distribution system that is evolvable to support continuous robotic/human operation and is capable of scaling to global power utilization and industrial power levels." [131]

Nor have U.S. efforts been limited to NASA and private industries. Though amounting to an invisible sliver (~1%) of the $26 Billion annual Department of Defense Space budget, the DoD has put its toes into the water. Several articles and reports by DoD authors anticipate the importance of a Cislunar economy,[132] and NASA and the U.S. Space Force (USSF) have a Memorandum of Understanding for cooperation which covers "capabilities and practices enabling safe, sustained cislunar operations such as communications; navigation; space structure servicing, assembly,

---

128. White House and NSTC, "In-Space Servicing, Assembly, and Manufacturing National Strategy" (Executive Office of the President of the United States, April 2022), https://www.whitehouse.gov/wp-content/uploads/2022/04/04-2022-ISAM-National-Strategy-Final.pdf.
129. Tim Fernholz, "The US Has a National Strategy to Put Factories in Space," Quartz, June 16, 2022, https://qz.com/2163811/us-plans-tech-to-refuel-satellites-and-build-factories-in-space/.
130. NASA, "MOON TO MARS OBJECTIVES," September 2022, https://www.nasa.gov/sites/default/files/atoms/files/m2m-objectives-exec-summary.pdf.
131. NASA.
132. David Buehler et al., "Posturing Space Forces for Operations Beyond GEO"; Olson et al., "SSIB 2021."

and manufacturing and interoperability" as well as space logistical supply & support, and search rescue and recovery of human spaceflight.[133]

Well within a decade, the USSF hopes to orbit a Cislunar Highway Patrol (CHPS) satellite in the vicinity of the Moon for Space Domain Awareness (SDA).[134] The Defense Advanced Research Projects Agency (DARPA), the agency that brought you the Internet, GPS, and stealth is already exploring how Lunar materials might be used to construct Optical, Radio-Frequency, and Solar Panels in space through its Novel Orbital and Moon Manufacturing Materials and Mass Efficient Design (NOM4D) program.[135] DARPA is also developing a high-speed nuclear thermal rocket due to launch by 2026[136] called the Demonstration Rocket for Agile Cislunar Operations (DRACO).[137]

Most who have ever considered 'the strategic importance of the Moon' likely consider it a long-term concern. However, as detailed above, significant developments will mature in the next 10 years,

---

133. NASA and USSF, "NASA - U.S. Space Force Memorandum of Understanding," NASA, September 21, 2020, https://www.nasa.gov/sites/default/files/atoms/files/nasa_ussf_mou_21_sep_20.pdf.
134. David Buehler, "Cislunar Highway Patrol System (CHPS)," January 7, 2021, https://www.afrl.af.mil/News/Photos/igphoto/2002556344/mediaid/4752579/.
135. DARPA, "Orbital Construction: DARPA Pursues Plan for Robust Manufacturing in Space," February 5, 2021, https://www.darpa.mil/news-events/2021-02-05.
136. American Nuclear Society, "DOD Seeks In-Space Demo of Nuclear Rocket Engine in FY 2026," Nuclear Newswire, May 9, 2022, https://www.ans.org/news/article-3946/dod-seeks-inspace-demo-of-nuclear-rocket-engine-in-fy-2026/.
137. Nathan Greiner and Tabitha Dodson, "Demonstration Rocket for Agile Cislunar Operations," DARPA, August 2022, https://www.darpa.mil/program/demonstration-rocket-for-agile-cislunar-operations.

well inside the typical length of time to specify, design and acquire major space systems.

So What? Of What Consequence are Civil / Commercial Lunar Activities to USSF Plans?

Of what concern is it to the Space Force that there is any level of civil or commercial activity on or near the Moon? Isn't the Moon protected by the Outer Space Treaty for purely peaceful purposes? Can't NASA take care of itself? Don't commercial actors assume their own risk? Doesn't the Space Force have more important, pressing concerns supporting the Army, Navy and Air Force in the Indo-Pacific, Europe, and Middle East?

The USSF and USSPACECOM certainly have such obligations, and certainly must focus the bulk of their resources on these extant challenges. But the Unified Command Plan (UCP) assigns to USSPACECOM the entirety of the universe above 100km as an Area of Responsibility (AOR)[138] equivalent to the AOR responsibilities of EUCOM for EUROPE, AFRICOM for AFRICA, CENTCOM for Central Asia / Middle-East, and INDOPACOM for the Indo-Pacific. The UCP also tasks them with defense of commerce. [139] Likewise, the USSF is legally assigned the function of protecting America's interests in the space domain.[140] Former NASA administrator Bridenstine, while still a U.S. representative

---

138. James H. Dickinson, "Fiscal Year 2023 Priorities and Posture of United States Space Command," Senate.gov, March 1, 2022, https://www.armed-services.senate.gov/imo/media/doc/USSPACECOM%20FY23%20Posture%20Statement%20SASC%20FINAL.pdf.
139. James H. Dickinson.
140. U.S. Congress, "PUBLIC LAW 116–92—DEC. 20, 2019: NATIONAL DEFENSE AUTHORIZATION ACT FOR FISCAL YEAR 2020," December 20, 2019, https://www.govinfo.gov/content/pkg/PLAW-116publ92/pdf/PLAW-116publ92.pdf.

stated, "It must be stated that constitutionally, the U.S. government is required to provide for the common defense. This includes defending American military assets in space AND commercial assets in space, many of which have and will have a dual role of providing commercial and military capabilities…to enable freedom of action, the United States must have cis-lunar situational awareness, a cis-lunar presence, and eventually must be able to enforce the law through cis-lunar power projection." [141]

141. Jim Bridenstine, "This Is Our Sputnik Moment.," *OKGrassroots* (blog), November 7, 2016, https://okgrassroots.com/?p=642815.

Appendix: Dominant Resource Strategies

Dominant 'Looking Down' Strategies

What are the right moves and trades below GEO (B-GEO)? The USSF must move toward a more resilient architecture and free sufficient resources for Space Control and Tactical ISR (including space-based radar to track moving targets). The obvious moves are to use the current approach of Space Development Agency (SDA) to make use of rapid-turnover and large constellations in LEO (proliferated LEO), to move to a "Hybrid Architecture" incorporating commercial and allied capabilities across a diversity of orbits, and to purchase commercial capabilities (imagery, SATCOM, data).

Two other opportunities present themselves. First, the USSF could posture itself to take advantage of Starship-enabled experimentation. If successful, Starship will enable 100MT to Low Earth Orbit (LEO) for a cost of $10m per sortie, or just $100/kg, which is approximately double the payload, and 1/23rd the cost of the best available option.[142] Volume also matters, and Starship volume is 1,000 m^3 nearly three times as much as Delta IV Heavy and nearly five times the volume of Falcon 9 / Falcon Heavy.[143] Further, while the fastest launch tempo was last year when SpaceX launched on average once every 12 days, Starship is designed to launch 3x per day.[144] Essentially for the same launch budget, the USSF could conduct 35x the current level of experimentation, and would need to posture itself to build prototypes at such a scale.

---

142. Peter Garretson and Samuel Havard, "The Starship Singularity."
143. Peter Garretson and Samuel Havard.
144. Peter Garretson and Samuel Havard.

Second, military airlift provides support to a diversity of users, who pay their own way via a Working Capital Fund called TWCF.[145] The Space Force should proactively seek to create a Working Capital Fund for space services it provides to the other military services (Army, Navy, Air Force, Marines, Coast Guard) and intelligence community (IC).

Dominant 'Looking Out' Strategies

What are the right moves and trades beyond GEO (X-GEO)? The key is to posture ahead to "be there." Establishing basic presence and basic infrastructure services for Space Security (SDA, PNT, MILSATCOM) could be done by extending the SDA "Tranche" smallsat model for a low-cost military XGEO constellation.[146] A complementary approach is to purchase such services or data from commercial providers who are already developing Cislunar business plans. This might include the lease of contractor owned, contractor operated (COCO) facilities which facilitate XGEO logistics and presence. It also makes sense to leverage capabilities in tandem with the larger commercial investment, including the purchase of commercial transportation services, commodities (such as space-sourced propellants, structural and functional materials) that

145. U.S. Air Force, "UNITED STATES AIR FORCE WORKING CAPITAL FUND (Appropriation: 4930) Fiscal Year (FY) 2024 Budget Estimates," March 2023, https://www.saffm.hq.af.mil/Portals/84/documents/FY24/Working%20Capitol%20Fund/24PB%20Congressional%20Final.pdf?ver=BCaiH27Dxlv_nz84wZS1Xw%3D%3D.

146. DoD, "Space Development Agency Makes Awards for 126 Satellites to Build Tranche 1 Transport Layer," U.S. Department of Defense, February 28, 2022, https://www.defense.gov/News/Releases/Release/Article/2948229/space-development-agency-makes-awards-for-126-satellites-to-build-tranche-1-tra/https%3A%2F%2Fwww.defense.gov%2FNews%2FReleases%2FRelease%2FArticle%2F2948229%2Fspace-development-agency-makes-awards-for-126-satellites-to-build-tranche-1-tra%2F.

establish the basics of a Cislunar logistical system, including propellant, refueling, servicing. Options to maximize capability while minimizing cost would include establishing a Civil Reserve Space Fleet (analogous to the Civil Reserve Air Fleet (CRAF) or Merchant Marine),[147] or potentially leasing mobility assets (for example, a flexible, low-cost asset like Starship). Finally, science and technology investments need to focus on those elements which are logistics and force multipliers: In-situ Resource Utilization (ISRU), In-Space Servicing Assembly and Manufacturing (ISAM), power and propulsion.

One expression of this might encourage an initial XGEO experimentation campaign on the order of $200M, followed by a "Tranche 1" of Cislunar infrastructure of basic SDA, PNT and MILSATCOM, perhaps $5B to deploy ~100 satellites over 5 years, and some number of leased high mobility "patrollers" which traverse the Cislunar lines of commerce, and expand in proportion to commerce.

---

147. Theresa Hitchens, "Space Force Seeks Ideas for Potential 'commercial Space Reserve' Fleet," *Breaking Defense* (blog), February 9, 2023, https://breakingdefense.sites.breakingmedia.com/2023/02/space-force-seeks-ideas-for-potential-commercial-space-reserve-fleet/.

PART II

China Space Ambitions: Drivers, Goals, and Capacities

Namrata Goswami

Abstract

Assessing China's space program is vital to understand how investments in space adds to a country's comprehensive national power. In analyzing China's space program, its goals, and its space capacities, it becomes evident that the Communist Party of China (CPC) leadership has prioritized investments in space as a key area of innovation and technology. China has both a robust civilian and military space structure, inspired and institutionalized by a civil-military integration strategy, with the aim of emerging as a lead actor in space by 2049, the hundredth-year celebration of the establishment of the People's Republic of China. This paper explains the drivers that inform China's space program, highlights its space goals and provides an assessment of China's space capacities (civilian, military, commercial). The paper ends by drawing out some scenarios and conclusions.

China views space as part of its grand strategy (comprehensive national power) and as critical for its national rejuvenation. First and foremost, China's space program is informed by its historical lessons to include avoiding a repeat of the 'century of humiliation' when China was caught off-guard due to its strategic blindness in recognizing the critical importance of defending its maritime and land borders, against external invasion.[148] Consequently, maintaining freedom of access to its space lines of communication is recognized as a vital national security requirement.

In its White Paper titled "China's Space Program: A 2021 Perspective," China stated that "the space industry is a critical element of the overall national strategy";[149] that the mission of China's space program is to "protect China's national rights and interests, and build up its overall strength." That the core principle of China's space program is that "China's space industry is subject to and serves the overall national strategy".[150] Critically, the 2021 White paper aims to further develop space infrastructure within the next five years (2021-2026) to include 'remote sensing satellites, the BeiDou navigation system, human spaceflight, dual-antenna X-band interferometric synthetic aperture radar (InSAR), lunar and asteroid probes, planetary defense, nuclear propulsion, in orbit service and maintenance, reusable space transport systems.[151] Since 2017, President Xi Jinping has included his idea of a 'China Dream' of national rejuvenation into the Chinese constitution,[152] and space forms an integral part of that China Dream.[153] In April 2020, the

---

148. Jun. J. Nohara, "Sea Power as a Dominant Paradigm: The Rise of China's New Strategic Identity", *Journal of Contemporary East Asian Studies*, Vol.6, No.2 (2017): 210-232, https://www.tandfonline.com/action/showCitFormats?doi=10.1080%2F24761028.2017.1391623

149. China National Space Administration, "China's Space Program: A 2021 Perspective", http://www.cnsa.gov.cn/english/n6465652/n6465653/c6813088/content.html (Accessed March 24, 2022).

150. Ibid.

151. Ibid.

152. Constitution of the Communist Party of China, Revised and adopted at the 19th National Congress of the Communist Party of China on October 24, 2017, p.3, http://www.xinhuanet.com/english/download/Constitution_of_the_Communist_Party_of_China.pdf (Accessed March 31, 2022).

153. "Full text of Xi Jinping's Report at 19th CPC National Congress," Xinhua, November 04, 2017,

powerful China National and Reform Commission identified space as 'new infrastructure" thereby prioritizing investments in space capacity building propelled by Xi's Innovation and 'Made in China' policies.[154]

Funded annually with a budget of approximately $11billion based on open source data [China does not publish an official space budget],[155] China has ambitions to establish a permanent space station (2022);[156] a lunar research outpost (2036),[157] a Solar Power Satellite (SPS) transmission capacity from Geosynchronous Orbit (GEO) by 2050,[158] and establish itself as the world leader in space by 2049, the hundredth-year celebration of the establishment of the People's Republic of China (PRC).[159] Besides civilian space ambitions, China has invested in developing its military space capacity specially to augment its informatization warfare capacity

---

https://www.chinadaily.com.cn/china/19thcpcnationalcongress/2017-11/04/content_34115212.htm (Accessed March 24, 2022).

154. "Xi Stresses Innovation as Primary Driving Force for Development", Xinhua, October 30, 2020, http://www.xinhuanet.com/english/2020-10/30/c_139478353.htm (Accessed March 25, 2022).

155. Chethan Kumar, "Space Race: China in Orbit with 7 Times More Funding", *The Times of India*, February 1, 2020, https://timesofindia.indiatimes.com/business/india-business/space-race-china-in-orbit-with-7-times-more-funding/articleshow/73824661.cms (Accessed March 25, 2022).

156. Huaxia, "China's Space Station to Host 6 Astronauts by end of 2022", Xinhua, March 04, 2022, http://www.xinhuanet.com/english/20220304/b2d55553be114fa4b7bfc6a5c2f83e03/c.html (Accessed March 25, 2022).

157. "China unveils Follow-Up Lunar Exploration Missions", *Xinhua*, January 14, 2019, http://www.xinhuanet.com/english/2019-01/14/c_137743015.htm (Accessed March 24, 2022).

158. Zhao Lei, "Scientists Envision Solar Power Station in Space", *China Daily*, February 27, 2019, http://www.chinadaily.com.cn/a/201902/27/WS5c75c8b3a3106c65c34eb8e3.html (Accessed March 25, 2022).

159. Namrata Goswami, "China in Space: Ambitions and Possible Conflict," *Strategic Studies Quarterly* (Spring 2018): 74-97. https://www.airuniversity.af.edu/Portals/10/SSQ/documents/Volume-12_Issue-1/Goswami.pdf (Accessed March 31, 2022).

for Military Command, Control and Communication (CnC), Positioning, Navigation and Timing (PNT) for independent missile launch and tracking as well as sea-based launch for avoiding detection. In the commercial sphere, Chinese private space companies like OneSpace have already launched to orbital space in 2019 and has plans for developing reusable rockets, now a critical goal in the next five years (2021-2026) as per its 2021 White Paper on its space activities. China has invested up to $3 billion in its private space sector annually since 2018.[160] As per its State Council released Document titled "Guiding Opinions of the State Council on Innovating the Investment and Financing Mechanisms in Key Areas and Encouraging Social Investment",[161] otherwise known as Documents 60 (2014), encouraging investments in the private space sector and help subsidize innovation in key space technologies to gain strategic advantage is a key goal for China. Space forms an integral part of its Belt and Road Initiative (BRI) Spatial Information Corridor since 2018.[162] China has utilized policies to direct and rein in its own private space sector under its Civil-Military Fusion Strategy and its 2021 National Defense Law. This law is legally binding and states that

> All State bodies, armed forces, political parties, people's organizations, enterprises, public institutions, social organizations, and other organizations shall support and

160. Jeff Foust, "Space Investments hits Record High in 2018", *Spacenews*, April 10, 2019, https://spacenews.com/space-investment-hits-record-high-in-2018/ (Accessed March 31, 2022).
161. China Public Private Partnership Center, "Guiding Opinions of the State Council on Innovating the Investment and Financing Mechanisms in Key Areas and Encouraging Social Investment) was the start of China's modern commercial space sector", April 14, 2015, https://www.cpppc.org/en/zy/994006.jhtml (Accessed March 24, 2022).
162. "China to Boost Int'l Cooperation on Space Exploration", *Xinhua*, June 6, 2021, http://www.xinhuanet.com/english/2021-06/12/c_1310004733.htm (Accessed March 31, 2022). For those interested in an in-depth assessment, *see Min Ye, The Belt Road and Beyond: State-Mobilized Globalization in China, 1998–2018* (Cambridge: Cambridge University Press, 2020).

participate in accordance with the law in national defense development, and fulfill their national defense duties and tasks… Any organization or individual who, in violation of this Law or other relevant laws, refuses to fulfill their national defense obligations or jeopardizes national defense interests shall be held legally responsible.[163]

In light of this backdrop, this paper highlights the key drivers of China's space program, the goals of China's space program, and its civilian, military and commercial capacities. The paper ends by drawing out some scenarios and conclusions.

Key Drivers of China's Space Program

The key drivers of China's space program are assuming leadership in space; regime legitimacy and internal national development; economic development of space and entrepreneurship; national security.

---

163. Law of the People's Republic of China on National Defense Order of the President of the People's Republic of China, December 26, 2020, came into effect January 2021, http://www.npc.gov.cn/englishnpc/c23934/202109/567129ffe3144ccb9ff358fed798b9e3.shtml (Accessed March 31, 2022); for those interested in reading the Chinese version, please see 中华人民共和国国防法（1997年3月14日第八届全国人民代表大会第五次会议通过根据2009年8月27日第十一届全国人民代表大会常务委员会第十次《关于修改部分法律的决定》修正2020年12月26日第十三届全国人民代表大会常务委员会第二十四次会议修订），http://www.mod.gov.cn/regulatory/2020-12/27/content_4876050.htm (Accessed March 31, 2022).

1. Leadership in Space

For China, leadership in space is critical as it grants it the ability to constitute the norms, rules and regulations of space governance, as well as guarantees its citizens the ability to access space for its economic prospects. In line with this driver, China's space behavior and strategy is informed by its historical texts, to include Han Feizi and his focus on legalism, an assertive and strong leader, and a strong state;[164] Sun Tzu's *Art of War* and his focus on a comprehensive legitimate grand strategy;[165] and Mao Tse Tung's *On Guerilla Warfare* on how to gain the initiative by utilizing both engagement and disengagement of the adversary (space warfare).[166] President Xi Jinping has offered us his concept of Chinese leadership in his China's dream,[167] inscribed into the Communist Party of China (CPC) constitution in 2017.[168] In his communications with China's space scientists, Xi pointed out that China's dream was to become an aerospace power with the space dream aimed at strengthening China's role in the world.[169] China has set itself a goal to exceed all others in space by 2049 in time for its 100-year celebration of the establishment of the PRC. The driver for leadership is aimed at developing comprehensive power: Industrial,

---

164. Stanford Encyclopedia of Philosophy. "Legalism in Chinese Philosophy", n.d. https://plato.stanford.edu/entries/chinese-legalism/ (Accessed March 25, 2022).
165. Sun Tzu, *The Art of War*, http://classics.mit.edu/Tzu/artwar.html (Accessed March 25, 2022).
166. Mao Tse Tung on Guerilla Warfare (Martino Fine Books, 2017).
167. China Dream, https://www.chinadaily.com.cn/china/Chinese-dream.html (Accessed March 25, 2022).
168. The State Council, People's Republic of China, "Full text of resolution on amendment to CPC Constitution," October 24, 2017, http://english.www.gov.cn/news/top_news/2017/10/24/content_281475919837140.htm (Accessed March 25, 2022).
169. Shi Yu, "On Chang'e 5 Mission Day, Looking Back on Xi's Key Quotes on China's Space Dream", CGTN, November 24, 2020, https://news.cgtn.com/news/2020-11-24/No-end-for-space-exploration-Xi-s-key-quotes-eye-China-s-space-dream-VGy11L9qEw/index.html (Accessed March 25, 2022).

Logistic, Diplomatic, Economic.[170] The goals, articulated by the China Academy of Launch Vehicle Technology (CALVT) was published in the front pages of the *People's Daily* in November 2017. [171] To classify China space program an exploration program is misleading; this is a program for industrial and economic dominance of the cislunar (space between Earth and Moon) System. As part of constituting leadership in space, China established the BRI spatial information corridor in 2018 offering its BeiDou navigation system and other space infrastructure to the 139 partner nations of the BRI.[172] The BRI includes several U.S. allies like Greece, Italy, Saudi Arabia, UAE, New Zealand., and several NATO members like Poland, Latvia, Lithuania, Estonia, Czech Republic, Bulgaria, Slovakia, and Romania.

2. Regime Legitimacy and Internal National Development

The CPC cares about regime legitimacy, both internally and internationally. Space plays an integral part in augmenting CPC regime legitimacy, by showcasing China's success under CPC guidance to explore and develop space. Successful missions to the

---

170. See Stacey Solomone, *China's Strategy in Space* (New York: Springer, 2013). Also see Brian Harvey, *China's Space Program-From Conception to Manned Spaceflight* (New York: Springer Praxis Books, 2004). Also see Michael Pillsbury, *The Hundred-Year Marathon China's Secret Strategy to Replace America as the Global Superpower* (New York: St Martin's Griffin, 2016).

171. Stephen Chen, "China's Nuclear Spaceships will be 'Mining Asteroids and Flying Tourists' as it aims to overtake the US in a Space Race," *South China Morning Post*, November 17, 2017 at https://www.scmp.com/news/china/policies-politics/article/2120425/chinas-nuclear-spaceships-will-be-mining-asteroids (Accessed March 31, 2022).

172. David Sacks, "Countries in China's Belt and Road Initiative: Who's In and Who's Out", Council on Foreign Relations, March 24, 2021, https://www.cfr.org/blog/countries-chinas-belt-and-road-initiative-whos-and-whos-out (Accessed March 25, 2022).

Moon and Mars, lunar sample return, the successful launch of *Tianhe*, the cargo spaceship *Tianzhou*, the world's first Quantum satellite, Mecius, the Tiangong space station, developing the BeiDou navigation system, all add to the CPC legitimacy as a force for innovation in high end technology. The centrality of Xi and his thought published as manual is now part of China's school curriculum, highlighting 14 principles and the centrality of the CPC in developing China into a first-rate power and a leader in science and technology.[173]

China's space program is directly linked to internal national development since its start in 1956. Space is utilized for navigation, e-commerce, ATM transactions, disaster relief, weather prediction for agriculture, fishery, e-education, telemedicine, tele-education, satellite television and satellite internet. Almost all of China's White Papers on space highlight the critical connection between space and comprehensive national development.[174] China views space as an infrastructure, critical for internal national development and international recognition as a great power in space.

3. Economic Development of Space and Entrepreneurship

The economic dimension of space and its future potential is highlighted as critical for China's space program. The Earth-Moon zone is predicted to generate around $10 trillion annually by

173. "Xi Jinping Thought Added into Curriculum: Ministry of Education", *The Global Times*, August 24, 2021, https://www.globaltimes.cn/page/202108/1232364.shtml (Accessed March 25, 2022).
174. State Council, People's Republic of China, "China's Space Activities", November 2000, https://www.fmprc.gov.cn/ce/celt/eng/zt/zfbps/t125258.htm (Accessed March 25, 2022).

2050.[175] The BeiDou navigation system, established with an investment of $9billion is predicted to generate $59 billion in revenue for China annually.[176] Researcher, Li Mingtao, with the National Space Science Center indicates that China's investment in space mining might become a new engine for the global economy.[177]

Since 2016, China is creating the avenues for encouraging its private space sector and subsidizing entrepreneurship. As stated earlier, clear direction has been offered for the private space sector via China's Document 60. Moreover, subsidized rates of access to spaceports maintained by the People's Liberation Army (PLA) for Chinese private launches is offered plus the financial investments in developing 5G satellite networks as well as reusable launch systems. Chinese private space companies are building small satellites, working on creating a 5G satellite constellation in Low Earth Orbit (LEO), and investing in commercial space launch sites.

4. National Security

China awoke to the national security aspect of space during the First Gulf War of 1991 as it watched U.S. space-based support "coordinate forces beyond 'line of sight', target and track integrated air defense system, and use its Global Positioning System (GPS) for

---

175. Cao Siqi, "China Mulls $10 trillion Earth-Moon Economic Zone", *The Global Times*, November 1, 2019, https://www.globaltimes.cn/page/201911/1168698.shtml (Accessed March 25, 2022).
176. Mark Andrews, "Navigating the World", CKGSB Knowledge, March 15, 2021, https://english.ckgsb.edu.cn/knowledges/china-global-navigation-system-beidou/ (Accessed March 25, 2022).
177. "China Focus: Capture an Asteroid, bring it back to Earth?", Xinhuanet, July 23, 2018, http://www.xinhuanet.com/english/2018-07/23/c_137342866.htm, (Accessed March 25, 2022).

precision strike, and anticipate when missiles were incoming."[178] China suddenly understood the inefficacy of its own surface to air missile defense system in light of U.S. space capacities. The 1996 Taiwan Strait crisis was seen as a humiliation by China when it lost sight of two of its missiles over the Taiwan Strait allegedly due to the U.S. cutting off GPS signal over the Pacific.[179] Consequently, China developed its own version of GPS-the BeiDou navigation system. China has now strengthened the national security dimensions of space by focusing on developing overhead sensing, long term over the horizon communication, missile warning and tracking, Anti-Satellite (ASAT) weapons, Rendezvous and Proximity Operations (RPOs), blinding and dazzling, a dual use robotic arm capacity and an independent capacity to guide its own missiles. The Political Bureau of the CPC Central Committee established the Central Commission for Integrated Military and Civilian Development (CCIMCD) in January 2017, the first such body dedicated to civil-military integration to be established by the top CPC leadership.[180] In March 2018, Xi presided over a plenary

---

178. Namrata Goswami, "Asia's Space Race: China Leads India on Strategy", *The Interpreter*, March 9, 2022, https://www.lowyinstitute.org/the-interpreter/asia-s-space-race-china-leads-india-strategy (Accessed March 31, 2022). Also see Harlan W. Jencks, "Chinese Evaluations Of "Desert Storm": Implications for PRC Security", *The Journal of East Asian Affairs*, Vol. 6, No. 2 (Summer/Fall 1992), pp. 447-477.
179. Minnie Chan, "Unforgettable humiliation led to Development of GPS Equivalent", *South China Morning Post*, November 13, 2009, https://www.scmp.com/article/698161/unforgettable-humiliation-led-development-gps-equivalent (Accessed March 25, 2022).
180. Zhao Lei, "Civil-Military Integration will deepen", *China Daily*, March 3, 2018 at http://www.chinadaily.com.cn/a/201803/03/WS5a99d67ca3106e7dcc13f437.html (Accessed March 31, 2022). "Xi Stresses Military-Civilian Integration in a New Era", Speech Delivered by Xi Jinping at the First Session of the 13th People's Congress", *Xinhua*, March 21, 2018, https://www.chinadailyhk.com/articles/184/187/127/1521628772832.html (Accessed on March 20, 2022). Backgrounder: Xi Jinping's Vision for China's Space Development", Xinhua, April 24, 2017,

session of the CCIMCD.[181] In this session, Xi urged private companies across sectors (to include space private startups), to locate their work within this civil-military integration strategy, in order to build integrated national strategic systems and capabilities.

Keeping these four drivers, leadership in space; regime legitimacy and internal national development; economic development of space and entrepreneurship; national security, let us now assess China's space goals and its space capacities.

China's Space Goals

China supports and funds one of the most ambitious space programs on the globe. Viewing space investments as part of a larger infrastructure and logistical industrial development, China points out that its space investments will build resilient national systems that will help develop China nationally and internationally. Amongst the foremost space goals for China are to build its own permanent space station (the Tiangong by 2022), construct space based solar power satellites (SBSP) in Geosynchronous Orbit (GEO) by 2050, develop its lunar robotic mission program and establish a lunar station on the Moon by 2036, build its Mars robotic missions including Mars sample return by 2030 and develop a human landing system for Mars by 2049, asteroid probes (2029 to 2034), and develop planetary defense (2021 White paper). China also aims to build a resilient space support structure with a fleet of satellite constellations by 2030, that includes both civilian and military satellites.

---

http://eng.chinamil.com.cn/view/2019-02/18/content_9429066.htm (Accessed on March 06, 2022).
181. Ibid.

a. Permanent Space Station

Foremost in this space logistics and infrastructure building for China is to develop its own permanent space station called *Tiangong*. On April 29, 2021, China launched the core lab of the Tiangong, called the *Tianhe*. The *Tianhe*, now in its orbital slot in LEO between 350kms to 450 kms above Earth, is the management module of the *Tiangong*, with the ability to dock three spacecrafts for short periods of time. The *Tianhe* is designed for a lifespan of 10 years. The *Tianhe* is the largest spacecraft that China has launched till date. According to Bai Linhou, the Deputy Chief Designer of *Tiangong* at the China Academy of Space Technology (CAST) and under China Aerospace Science and Technology Corporation (CASC), "we will learn how to assemble, operate and maintain large spacecraft in orbit, and we aim to build Tiangong into a state-level space lab supporting the long stay of astronauts and large-scale scientific, technological and application experiments".[182] Linhou further elaborated that China's space station will build the capability for the utilization of space resources as well. In June 2021, China launched its first crewed mission of three astronauts to the *Tianhe*, for a period of three months, to help construct the space station.[183] The astronauts also tested the new extravehicular suits. Critically, China's cargo spacecraft, the *Tianzhou-2* launched supplies to the *Tianhe*, as part of the logistic system. When it is finally completed, the permanent space station, *Tiangong* will have its core module *Tianhe*, as well as the two labs, the *Wentian* and the second lab called the *Mentiang*. *Wentian* will be used for scientific experiments as well as offer living space, and will have a robotic arm to carry out extravehicular activities; the *Mentiang* will be similar in function

182. "China Launches Space Station Core Module Tianhe", *Xinhuanet*, April 04, 2021, http://www.xinhuanet.com/english/2021-04/29/c_139914454.htm (Accessed March 28, 2022).

183. "China Launches First Crewed Mission to for Space Station Construction", *Xinhua*, June 17, 2021, http://www.xinhuanet.com/english/2021-06/17/c_1310012945.htm (Accessed March 28, 2022).

and make. The *Tiangong* is a T-shaped space station with a maximum weight of about 66 tons and will possess the capability to support up to six astronauts for a period of six months stay, the longest such presence in space by Chinese astronauts.[184] Critically, China is launching a space telescope called *Xuntian*, that will orbit close to the space station, and help determine observations of space. Lin Xiqiang, the Deputy Director of the China Manned Space Agency (CMSA) specifies that 12 space launches will be required to complete the space station, and help construct it. China is testing a new renewable life support system as part of the space station. The station is geared by electric propulsion and has two flexible solar panels. The main experiments that Chinese space scientists will conduct are to study how humans can survive for longer periods in space. As part of China's 14th Five Year Plan (2021–25), the National Natural Science Foundation of China allocated $2.3 million for Chinese scientists to study the feasibility of in orbit assembly of a kilometer-wide space station.[185] The thrust of this particular project is to bring down the weight of construction materials to be launched to space, so that more construction materials can be launched at lower cost.

b. China's Lunar Exploration Program

On January 14, 2019, a few days after China successfully landed humanity's first robotic probe on the far side of the Moon with its *Chang'e 4* mission, Wu Yanhua, Deputy head of CNSA announced that by 2019 year-end, China will launch *Chang'e-5* to bring lunar

184. "Core Focus: More Details of China's Space Station Unveiled", Xinhua, May 18, 2020, http://www.xinhuanet.com/english/2020-05/18/c_139067133.htm (Accessed March 28, 2022).
185. Fan Anqi, "China to Study Assembly Mechanics of Kilometer-Level Extra-Large Spacecraft", *The Global Times*, August 25, 2021, https://www.globaltimes.cn/page/202108/1232426.shtml (Accessed March 28, 2022).

samples back to Earth; this will be followed by *Chang'e-6* (2024), aimed at bringing samples from the South Pole; *Chang'e-7* (2028) that will survey the South Pole for evaluating its composition. The *Chang'e-8* (2030) will test key technologies like 3D printing to lay groundwork for the construction of a scientific base on the moon (2036).[186] China signed a Memorandum of Understanding (MoU) with Russia to build an International Lunar Research Station on the Moon in March 2021.[187]

In December 2020, a year behind schedule but successful, China brought back samples from the Moon via its *Chang'e 5* mission and unfurled its first fabric flag on the Moon, a feat much lauded as establishing Chinese technology and leadership on the lunar surface. Wu Yanhua, in a special press conference at the State Council Information Office in Beijing stated that "The successful mission is a new milestone in the development of China's space industry, proving that China has mastered the technology for shuttling between Earth and the moon".[188] China's lunar program is propelled by resource prospecting visions. Scientists at the Technology and Engineering Center for Space Utilization of the Chinese Academy of Sciences (CAS) tested 3D printing technology in micro-gravity by successfully completing a ceramic testing technology in 2018.[189]

---

186. "China unveils Follow-Up Lunar Exploration Missions", *Xinhua*, January 14, 2019 , http://www.xinhuanet.com/english/2019-01/14/c_137743015.htm (Accessed March 31, 2022).
187. China National Space Administration, "China and Russia sign a Memorandum of Understanding Regarding Cooperation for the Construction of the International Lunar Research Station", March 09, 2021, http://www.cnsa.gov.cn/english/n6465652/n6465653/c6811380/content.html (Accessed March 28, 2022).
188. Deng Xiaoci and Fan Anqi, "Chang'e 5 Completes Lunar Trip, More Space Missions Planned", *The Global Times*, December 17, 2020, https://www.globaltimes.cn/page/202012/1210275.shtml (Accessed March 28, 2022).
189. "China Focus: China Pioneers Ceramic 3D Printing in Micro-Gravity", Xinhua, June 19, 2018 at http://www.xinhuanet.com/english/2018-06/19/c_137265536.htm (Accessed March 31, 2022). Richard P. Appelbaum,

According to Wang Gong, Director of the CAS Key Laboratory of Space Manufacturing Technology, this evaluated Chinese capability to build bases on the moon and Mars, as well as in-situ resource utilization and space manufacturing with space based resources.[190] Using ceramics is instructive as it is similar in composition to lunar silicate particles. As Wang put it, "Elon Musk and SpaceX are developing technologies to take people to other planets, and we are developing technologies to help them survive."[191] Wu Weiren, the chief scientist of CLEP. Stated in 2005 "our short-term goal is to orbit the Moon, and land on the Moon, and take samples back from the Moon…our long-term goal is explore, land and settle. We want our manned lunar landing to stay for longer periods and establish a research base,"[192] a goal now articulated with Russia. Weiren believes that the critical step forward is to establish a lunar palace especially on the South Pole by 2030, given the presence of sunlight and water-ice there. Beihang University in China has experimented with a simulated lunar module in 2018, in which eight students lived in a Moon lab for 370 days, to study how a regenerative life support system might work.[193] Significantly, the *Chang'e 4* carried a 3 kg, (0.8 liters) aluminum alloy cylinder, containing seeds of cotton,

---

et.al., *Innovation in China: Challenging the Global Science and Technology System* (Polity, 2018). George S. Yip and Bruce McKern, *China's Next Strategic Advantage:* From Imitation to Innovation (MIT Press, 2017).

190. "China Focus: China pioneers Ceramic 3D Printing in Microgravity", *China Daily*, June 19, 2018, http://www.chinadaily.com.cn/a/201806/19/WS5b291e1aa3104bcf48c155d9.html (Accessed March 31, 2022).

191. Ibid.

192. "What does China want to do in Space", BBC, April 20, 2016 at https://www.bbc.com/news/av/36089689/what-does-china-want-to-do-in-space (Accessed March 31, 2022). Marina Koren, "China's Growing Ambitions in Space", *The Atlantic*, January 23, 2017 at https://www.theatlantic.com/science/archive/2017/01/china-space/497846/ (Accessed March 31, 2022).

193. Guo Meiping, "China's 'Lunar Palace' Breaks Record after 365+5 days", CGTN, May 15, 2018, https://news.cgtn.com/news/3d3d514e7a67444e77457a6333566d54/share_p.html (Accessed March 31, 2022).

potatoes and Arabidopsis, a plant related to cabbage as well as eggs of silk worms.[194] The idea was to experiment if potatoes (food for Moon settlers), and silk worms can flourish on the Moon in a simulated mini biosphere. According to the mission chief, Liu Hanlong, the idea was to study the process of developing food for space travelers on the lunar surface. Liu specified, "Our experiment might help accumulate knowledge for building a lunar base and long-term residence on the Moon."[195] In May 2018, the *Queqiao* relay satellite, or Magpie Bridge, was placed in the Lagrange (L2) Halo orbit to serve as a communication relay satellite from the Chang'e 4.[196] *Queqiao* can peer into the Lunar polar craters, enabling future landing zones for China's probes in the shadowed regions, as part of recognizing the critical importance of L2 in cislunar space.

c. Space-Based Solar Power (SBSP)

China became the first country in the world to establish a state funded SBSP base plant in Chongqing's Bishan district, early this year.[197] The base plant is being constructed under the guidance of the Chongqing Collaborative Innovation Research Institute for Civil-Military Integration (CCIRICMI) in Southwestern China in partnerships with researchers from Chongqing University, CAST's

---

194. Rafi Letzter, "There are Plants and Animals on the Moon Now (because of China)", LiveScience, January 3, 2019 at https://www.livescience.com/64413-china-space-moon-plants-animals.html (Accessed March 31, 2022).
195. Ibid.
196. Luyuan Xu, " How China's Lunar Relay Satellite Arrived in its Final Orbit", Planetary Society, June 15, 2018, http://www.planetary.org/blogs/guest-blogs/2018/20180615-queqiao-orbit-explainer.html (Accessed March 31, 2022).
197. Zhao Lei, "Scientists Envision Solar Power Station in Space", *China Daily*, February 27, 2019,
http://www.chinadaily.com.cn/a/201902/27/WS5c75c8b3a3106c65c34eb8e3.html (Accessed March 31, 2022). For more on SBSP, see Brain O'Leary, Construction of Satellite Solar Power Stations from Non-Terrestrial Material", *Journal of Energy*, 1/3, 1977, pp. 155-158.

Xi'an Branch in Shaanxi province, and Xidian University. The initial investment for the SBSP plant of $15 million has been made by the Bishan district government. Technologies being tested include the construction of SBSP satellites in GEO using automated assembly and the wireless transmission of power.[198] Li Ming, senior vice president of CAST, asserted that China will lead the world in this critical renewable energy source.[199] The key challenges that the plant will be testing are:

a. Microwave transmission of electricity,
b. In-space manufacturing of SBSP satellites.

Xie Gengxin, Deputy Head of CCIRICMI stated:

> We plan to launch four to six tethered balloons from the testing base and connect them with each other to set up a network at an altitude of around 1,000 meters… these balloons will collect sunlight and convert solar energy to microwave before beaming it back to Earth. Receiving stations on the ground will convert such microwaves to electricity and distribute it to a grid… if everything goes well, a Chinese solar power station will be put into orbit about 36,000 kilometers above Earth and start generating power before 2040.[200]

Wang Xiji, the chief designer of China's first rocket, the *Long March 1* advocates for SBSP in China. Wang believes that "The world will panic when the fossil fuels can no longer sustain human development. We must acquire space solar power technology before

---

198. Pan Zhaoyi, "China Starts to Build World's First Space Based Solar Power Plant", CGTN, February 18, 2019, https://news.cgtn.com/news/3d3d514f34597a4e32457a6333566d54/index.html (Accessed March 31, 2022).
199. Ibid.
200. Ibid.

then…Whoever obtains the technology first could occupy the future energy market. So it's of great strategic significance."[201] Xiji further specified that "Once completed, the solar station, with a capacity of 100MW, would span at least one square kilometre, dwarfing the International Space Station and becoming the biggest man-made object in space".[202] Xiji believed that if China needs to act quickly on an SBSP program to ensure its ahead of the U.S. and Japan on this critical renewable technology and occupy strategically important locations in space.[203] The timeline for China's SBSP ambition is to test the first 100kW SBSP demonstration at LEO by 2025, 1 mW of SBSP power demonstration in GEO by 2030, demonstrate a 100 mW SBSP by 2035, construct a SBSP to orbit (36,000 kms) above Earth and start generating power by 2040 and establish the first commercial level SBSP in operation in GEO by 2050.[204]

d. Mars Mission

China officially launched its first independent Mars mission with an aim to orbit, land and sent out a rover to Mars surface in 2016. China's first attempt to send a Mars orbiter atop the Russian rocket Phobos-Grunt Spacecraft in 2011, failed to leave LEO and fell back

---

201. Chinese Scientists Mull Power Station in Space", *Space Daily*, March 31, 2015,
http://www.spacedaily.com/reports/Chinese_scientists_mull_power_station_in_space_999.html (Accessed March 31, 2022).
202. Ibid.
203. Ibid.
204. For elaboration on sources, please see Namrata Goswami, "Statement of Dr. Namrata Goswami
Independent Senior Analyst and Author 2016-2017 Minerva Grantee Before the U.S.-China Economic and Security Review Commission Hearing on "China in Space: A Strategic Competition?" April 25, 2019,
https://www.uscc.gov/sites/default/files/Namrata%20Goswami%20USCC%2025%20April.pdf (Accessed March 31, 2022).

onto the Pacific Ocean.[205] Four years later since establishing a Mars mission, China launched its first independent mission, the *Tianwen 1* to Mars in July 2020 and landed on the surface in February 2021, the second country after the U.S. to have a successful landing mission that communicated back to Earth. For China, the successful Mars mission is a process of building capacity to develop settlement for the most Earthlike planets. In 2019, China opened its first 53,330 square meters Mars simulation base (at a cost of $22.3 million) in Mangai city, located in the arid desert region of Qaidam Basin in the Qinghai-Tibet plateau, aimed at introducing people to life in a Mars-like environment. This region's barren, rocky landscape closely resembles the topographical conditions found on Mars.

Similar to the relay technology developed for China's lunar far side mission, China has developed relay capacity for Mars communication. The Mars landing also developed further China's automated landing as Mars atmosphere required a heat shield, parachute and thrusters to slow down the descent of the lander. Jia Yang, the deputy chief designer of the *Chang'e 3* probes specified that "the Mars rover should be able to sense the environment, plan its route, conduct scientific exploration and detect faults autonomously. It should be a mobile intelligence."[206] In fact, after it landed on Mars surface and sent out the rover in May 2021, the message from the Mars orbiter had a delay of 17 minutes to reach Earth due to the 320-million-km distance between Earth and Mars. With the Mars orbiting, landing and sending out the rover, China became the first country in the world to successfully achieve all

---

205. Leonard Davis, "Parts of Doomed Russian Mars Probe to Hit Earth in January", *Space.com*, December 27, 2011, https://www.space.com/14056-doomed-mars-probe-phobos-grunt-falling-debris.html (Accessed March 28, 2022).
206. "China Aims to Launch Mars Probe in 2020", *China Daily*, July 14, 2020, https://thediplomat.com/2020/07/why-is-china-going-to-mars/ (Accessed March 28, 2022).

three goals in its first independent attempt.[207] China plans to execute a Mars sample return, and establish a Mars human landing system and a Mars base by 2049. Wang Xiaojun, the head of the China Academy of Launch Vehicle Technology (CALT), in his presentation on "the Space Transportation System of Human Mars Exploration" highlighted some of the key technologies for Mars under consideration. Besides Mars sample return (2030), there are plans to develop robotic exploration of a site for a Mars base (2035), the human mission (2037), the building of a Mars base (2041) and the establishment of an Earth-Mars cargo system (2043). The technology that China views as enabling this Mars logistics is nuclear propulsion, something that China is working towards building. Wang referred to the logistics as a 'Sky ladder" system.[208] The roadmap for Mars was revealed by CALT.[209]

e. Asteroid Probes and Space Mining

China has announced that it would have an asteroid exploration mission by 2022. The aim of this mission will be to collect samples from the near-Earth asteroid Kamo'oalewa, and the same spacecraft, once it delivers the samples to Earth, will then be launched to the asteroid belt. This particular asteroid mission is being named after *Zheng He*, the ancient Chinese explorer and admiral of China's

---

207. "China Succeeds in First Mars landing", *Xinhua*, May 15, 2021, http://www.xinhuanet.com/english/2021-05/15/c_139947277.htm (Accessed March 28, 2022).
208. Deng Xiaoci, "Chinese Rocket Manufacturers Outlines manned Mars Mission Roadmap, Timetable", *The Global Times*, June 23, 2021, https://www.globaltimes.cn/page/202106/1226925.shtml (Accessed March 28, 2022).
209. "China Plans to Retrieve Samples from Mars by 2030, Study the Edge of the Solar System", *The Global Times*, March 07, 2022, https://www.globaltimes.cn/page/202203/1254136.shtml (Accessed March 28, 2022).

famous treasure ships fame in the 15th century.[210] The mission will utilize nanotechnology to carry out exploration of the Comet 133P/Elst–Pizarro. Critically, Russia has_joined in with China on this asteroid mission.[211] According to a China Academy of Space Technology (CAST) paper:

> ZhengHe flies an over ten-year mission, launching in 2022 to reach the NEA 2016HO3, return 200-1000g sample back to earth within 2-3 years, then continue its journey to rendezvous with the Main Belt comet 133P/Elst-Pizarro just before it reaches perihelion in 2030 and remain there for one year to carry on remote sensing and in-situ measurement.[212]

The second target would be comet 311P, a main comet in the Asteroid belt.[213] In its 2021 White Paper on space, China identified planetary defense as a critical area of study and leadership for China in the next five years (2021-2026). Chinese space scientists are working on plans to capture a Near Earth Asteroid (NEA) and bring it to Earth to inspect and extract its resources. Researcher, Li Mingtao, with the National Space Science Center under CAS along with his team details that plan.[214] The idea is for a spacecraft to bag

---

210. Xiaojing Zhang, et'al., "Zhenghe-A Mission to A Near-Earth Asteroid and A Main Belt Comet repl", n.d. https://www.hou.usra.edu/meetings/lpsc2019/pdf/1045.pdf (Accessed March 28, 2022).
211. Andrew Jones, "Russia Joins China's Mission to Sample an Asteroid and Study a Comet", Space.com, April 04, 2021, https://www.msn.com/en-xl/money/tech-and-science/russia-joins-chinas-mission-to-sample-an-asteroid-and-study-a-comet/ar-BB1fNps2?li=BBKxOg5 (Accessed March 28, 2022).
212. Ibid.
213. "China pushes forward exploration of small celestial bodies", *Xinhua*, April 24, 2021, http://www.xinhuanet.com/english/2021-04/24/c_139903587.htm (Accessed March 28, 2022).
214. "Capture an Asteroid, Bring it Back to Earth?", Chinese Academy of Sciences, June 24, 2018, http://english.cas.cn/newsroom/news/201807/t20180724_195396.shtml (Accessed March 31, 2022). For more on asteroid mining, please see John Lewis, *Mining the Sky Untold Riches from the Asteroids, Comets, and Planets*

an asteroid and push it over Earth, followed by a heat shield that unfolds reducing the velocity of the asteroid as it enters Earth's atmosphere. The landing must be minutely controlled so that it lands in an area far from human habitation. Li is working in collaboration with scientists from the Qian Xuesen Laboratory of Space Technology, under the China Aerospace Science and Technology Corporation (CASTC) to place satellites in the heliocentric Venus orbit, in order to search and analyze NEAs, with a diameter of 10 meters. The challenge after capture would be to drop the speed of the asteroid from 12.5 km per second to 140 meters per second before it touches down on its Earth landing spot. The timeline for such a launch to capture an asteroid is 2029 and the aim is to bring it back to Earth around 2034. A NEA like *3554 Amun*, approximately two kilometers in diameters, contains nickel and iron worth $8 trillion, cobalt ($6 trillion), other precious metals and gold worth $6 trillion, which together totals $20 trillion.[215] Based on such estimates, Li asserts that:

> Space mining might become a new engine for the global economy… Unlike missions to bring samples back, we aim to bring back a whole asteroid weighing several hundred tonnes, which could turn asteroids with a potential threat to Earth into usable resources…Our analysis shows that maneuvering a small asteroid is feasible in principle, and could bring enormous economic and social benefits.[216]

(New York: Helix Books, 1997). Robert Zubrin, *Entering Space: Creating a Spacefaring Civilization* (New York: TarcherPerigee, 2007).

215. Andrew M. Thorpe, *The Commercial Space Age: Conquering Space Through Commerce* (Bloomington: AuthorHouse, 2003), p. 149.

216. "China Focus: Capture an Asteroid, bring it back to Earth?", Xinhuanet, July 23, 2018, http://www.xinhuanet.com/english/2018-07/23/c_137342866.htm (Accessed March 31, 2022).

f. China's National Satellite Constellation

As per directions from the China National Development and Reform Commission in which the development of satellite internet is identified as 'new infrastructure', China has established a national plan to develop a national satellite mega-constellation of 12, 992 satellites as per filings with the International Telecommunications Union (ITU). The national satellite constellation is called "Guowang" or "national network" satellite internet project.[217] Bao Weimin, a senior member of the National Committee of the Chinese People's Political Consultative Conference and Director of the Science and Technology Committee of the Aerospace Science and Technology Group specified that a national grid company (Guowang) will be established to maintain this mega constellation.[218] This satellite internet project is supported by China's 14th Five Year plan (2021-26).

Given the drivers for China's space ambitions as well as its goals, what are some of China's space capacities today and what are the future space capacities that the country is prioritizing to achieve the space dominance it aspires to, by 2049.

China's Space Capacities

This section of the paper explains and discusses China's space capacities, to include its civilian, military and commercial space

---

217. Andrew Jones, "China is Developing Plans for a 13, 000 Satellite Megaconstellation", Space.com, April 21, 2021, https://spacenews.com/china-is-developing-plans-for-a-13000-satellite-communications-megaconstellation/ (Accessed March 28, 2022).

218. "Two Sessions | Bao Weimin: The state will set up a "State Grid" company to coordinate the construction and operation of satellite Internet!", March 10, 2021, https://www.163.com/dy/article/G4NC7RL705119QTF.html (Accessed March 28, 2022).

capacities and highlights some of the future capacities that China is working on.

Civilian Space Capabilities

China's civilian space capacity is led by its independent space launch system, its ability to maintain human presence in LEO, its missions to the Moon and Mars, and its independent BeiDou navigation system.

a. Independent Launch Capability

China possesses an advanced independent space launch capacity. Named the Long March (Chang Zheng) rocket series, the first *Long March-1* was based on the *Dong Feng 3* intermediate-range ballistic missile. The *Long March 1* launched China's first satellite into orbit in 1970. The *Long March 2* was based on the *Dong Feng 5* intercontinental ballistic missile (ICBM) and the *Long March 2F* launched China's first astronaut, Yang Liwei to space in 2003. Since 2016, China has developed a new series of launches called the Long March 5, 6 and 7. The Long March 5B has been specifically developed to launch the space station, as well as missions to the Moon. Wang Xiaojun, the head of CALT stated that the *Long March 5B* will expand China's aerospace capabilities. Compared to 14 tons to LEO by the *Long March 5*, the *Long March 5B* can carry 22 tons to LEO, and 14 metric tons to GEO. Taking 10 years to build the *Long March 5B*, the rocket's chief designer, Li Dong stated that "the development of the new carrier rocket also helps lay the foundation for the research and development of China's heavy-lift launch vehicle".[219] The development of launch systems is critical for meeting China's stated space goals. For instance, the successful development of the Long March 5B, that suffered two failed launched attempts,[220] until it finally succeeded, was the rocket that launched China's Mars mission and its *Chang'e 5* sample return mission. In June 2019, China also successfully launched the *Long March 11* from a mobile launch platform from the Yellow Sea off east China's Shandong Province. This is strategically important as it offers China a flexible platform to launch, brings down costs as its close to the equator, enables rapid responses in times of conflict, helps in stealth capabilities for launch as mobile sea space launches are difficult to find, and being a solid propellant rocket can launch quickly. This solid propellant capability is similar to the DF-21 Medium Range Ballistic Missile and DF-26 Intermediate Range Ballistic Missile.[221] Long Lehao, an academician of the Chinese Academy of Engineering and a chief designer at CALT, which builds the Long March rockets, specified what China's launch goals by 2030 are: "The capacity of Chinese rockets would reach 140 tonnes for low-Earth orbit, 44 tonnes for Earth-Mars transfer orbit, 50 tonnes for Earth-Moon transfer orbit, and 66 tonnes for geosynchronous transfer orbit in 2030." By 2030, China is developing the *Long March 9*, its super heavy lifter. The *Long March 9* is designed to carry a payload of 140 metric tons to LEO, a 50-ton spacecraft to a lunar transfer orbit and a 44-ton payload to Mars transfer orbit.[222] Importantly, in its justifications for the *Long*

---

219. "Long March-5B Enables China to Construct Space Station", *Xinhua*, May 05, 2020, http://www.xinhuanet.com/english/2020-05/05/c_139032959.htm (Accessed March 28, 2022).
220. Jeff Foust, "China's Powerful Long March 5 Rocket Fails on 2nd Launch", Space.com, July 02, 2017, https://www.space.com/37376-china-long-march-5-rocket-failure.html (Accessed March 28, 2022).
221. Namrata Goswami, "Why China's Long March 11 Launch Matters", *The Diplomat*, June 10, 2019, https://thediplomat.com/2019/06/why-chinas-long-march-11-launch-matters/ (Accessed March 28, 2022).
222. Zhao Lei, "Super Powerful Long March 9 Said to Begin Missions around 2030," *China Daily*, March 11, 2019.

March 9, China listed the following four goals: 1) launch a Mars robotic exploration mission which requires 41 tonne payloads, 2) manned Mars missions, 3) deep space missions, and 4) "Constructing orbital solar power plant with 10,000 MW capacity, massing some 50,000 tonnes, requiring 620+ launches)."[223]

China is also working towards developing reusability, a capability listed as of priority in the next five years in its 2021 White paper on space.[224] Amongst the experiments for reusability is the *Long March 8*, that launched in December 2020 with a capacity of 4.5 tons to sun synchronous orbit.[225] The *Long March 8* is viewed as a medium sized rocket, and the first stage of the *Long March 8* will become reusable by 2025. CASC, the developer of the *Long March 8* stated that by 2035, the entire rocket will become reusable. The *Long March 8* will be equipped to make changes to its flight path autonomously if it encounters malfunction.[226]

China has four spaceports. These include the Jiuquan Satellite Launch Center in the Gobi Desert, Inner Mongolia, the Taiyuan Satellite Launch Center in Kelan County, Xinzhou, Shanxi Province, the Wenchang Space Launch Site in Xichang, Sichuan, and the Xichang Satellite Launch Center in Xichang, Liangshan Yi

---

http://www.chinadaily.com.cn/a/201903/11/WS5c859b62a3106c65c34edcc0.html (Accessed March 28, 2022).

223. As quoted from Journal of Rocket Propulsion, January 2011, in https://www.globalsecurity.org/space/world/china/cz-x.htm (Accessed March 31, 2022).

224. See China National Space Administration, "China's Space Program: A 2021 Perspective", http://www.cnsa.gov.cn/english/n6465652/n6465653/c6813088/content.html (Accessed March 24, 2022).

225. "What's Special about China's Long March 8 Rocket?", *Xinhua*, December 23, 2020, http://www.xinhuanet.com/english/2020-12/23/c_139613492.htm (Accessed March 28, 2022).

226. Deng Xiaoci, "China's potentially reusable Long March-8 makes successful maiden flight", *The Global Times*, December 22, 2020, https://www.globaltimes.cn/page/202012/1210622.shtml (Accessed March 28, 2022).

Autonomous Prefecture in Sichuan. A fifth spaceport is under construction in Ningbo city, Zhejiang province. The spaceport will be equipped to support 100 launches in a year.[227] China is also building a ship 162.5 meters long with a 40-meter-width, termed a 'New type Rocket Launching Vessel',[228] capable of launching the *Long March 11* and other similar sized commercial rockets called the Smart Dragon series developed by China Rocket Co. Ltd affiliated with CALT. The Smart Dragon 3 will attempt its first flight in 2022. The ship is being built in the city of Haiyang, East China's Shandong Province.[229] This floating sea launch ship will also be utilized in the recovery of rockets.

BeiDou Navigation System

As on December 31, 2021, China had launched 499 satellites compared to the U.S. (2,944), Russia (169), India (118).[230] On June 23, 2020, China launched the 55th satellite of its BeiDou Navigation System (BDS) according itself a fully independent self-reliant Global Navigation Satellite System (GNSS) as an alternative to the U.S. Space Force-maintained Global Positioning System (GPS). With this launch, "China is now able to extend influence in a multidomain environment (land, sea and space) via its BeiDou

227. Reuters Staff, "China Begins Construction of its Fifth Rocket Launch Site", *Reuters*, April 07, 2021, https://www.reuters.com/article/us-space-exploration-china/china-begins-construction-of-its-fifth-rocket-launch-site-idUSKBN2BV0CF (Accessed March 28, 2022).
228. Deng Xiaoci, "China to deliver first rocket launching vessel by 2022", *The Global Times*, October 31, 2021, https://www.globaltimes.cn/page/202110/1237756.shtml (Accessed March 28, 2022).
229. Ibid.
230. Union of Concerned Scientists, "UCS Satellite Database," January 01, 2022, https://www.ucsusa.org/resources/satellite-database#:~:text=Assembled%20by%20experts%20at%20the,currently%20in%20orbit%20around%20Earth. (Accessed March 28, 2022).

space system, which provides navigation to aircraft, submarines, missiles, as well as commercial services dependent on such navigation".[231] This Chinese information infrastructure consists of undersea cables, where China is dominant, space-supported links, and other Earth-based links. BeiDou offers an alternative to GPS, enabling China to further consolidate its hold on global infrastructure and rulemaking, form partnerships and alliances and to control the standards for information technology, mobile devices, 5G, self-driving cars and drones, and the broader internet of things.[232] It offers China an influence mechanism where countries dependent on BeiDou would hesitate to criticize China on political issues such as Tibet, South China Sea (SCS), or Taiwan. This increases China's ability to coerce and compel. Wang Jingang, the chief deputy designer of the BDS-3 satellites specifies, "It's a rare chance to devote my intelligence to a symbolic national project…people still mainly depend on navigation by GPS, supplemented by BDS. I hope that in a few years, people can be navigated mainly by BDS".[233] BeiDou is now part of the BRI Spatial Information Corridor. Clearly, China's development of the BeiDou is a clear policy agenda of taking up leadership at the world of navigation and offer better services to countries that require it especially in the Asia-Pacific, Africa and Latin America.

---

231. Namrata Goswami, "The Economic and Military Impact of China's BeiDou Navigation System", *The Diplomat*, July 1, 2020, https://thediplomat.com/2020/07/the-economic-and-military-impact-of-chinas-beidou-navigation-system/#:~:text=By%20Namrata%20Goswami&text=On%20June%2023%2C%202020%2C%20China,for%20its%20BDS3%20navigation%20constellation (accessed March 25, 2022).

232. Eric Hagt, "China's BeiDou: Implications for the Individual and the State", SAIS Review, Vol. XXXIV, No. 1 (Winter-Spring 2014), pp. 129-140, https://www.jstor.org/stable/27000948 (Accessed March 28, 2022).

233. "From Compass to BeiDou: Chinese Wisdom Help Navigate Belt & Road", *China Daily*, November 19, 2018. https://global.chinadaily.com.cn/a/201811/19/WS5bf25cb3a310eff303289946.html (Accessed March 28, 2022).

Lunar and Deep Space Capacity

Wu Weiren, the chief scientist of CLEP states that "our short-term goal is to orbit the Moon, and land on the Moon, and take samples back from the Moon…our long-term goal is explore, land and settle. We want our manned lunar landing to stay for longer periods and establish a research base".[234] Lieutenant General Zhang Yulin, former Deputy Commander of China's Manned Space Program and former Deputy Chief of the Armament Development Department of the Central Military Commission (CMC), now with the Strategic Support Force (SSF) specified the significance of the Moon in 2016 when he stated, "the earth-moon space will be strategically important for the great rejuvenation of the Chinese nation".[235] Yulin indicated in that same interview that China will be investing in building capacity to generate solar power in space. He was clear on what China's space program focus should be in the long term, "The future of China's manned space program, is not a moon landing, which is quite simple, or even the manned Mars program which remains difficult, but continual exploration of the earth-moon space with ever developing technology".[236] Towards this end, China is developing capacity for operations in cislunar space as well as developing new spacecrafts that could take crewed missions to the Moon.

In a recent interview, Wu Weiren, the chief designer of China Lunar Exploration Program (CLEP) stated that China is now at Phase 4 of

---

234. "What does China want to do in Space", BBC, April 20, 2016 at https://www.bbc.com/news/av/36089689/what-does-china-want-to-do-in-space (Accessed on September 17, 2020). Marina Koren, "China's Growing Ambitions in Space", *The Atlantic*, January 23, 2017 at https://www.theatlantic.com/science/archive/2017/01/china-space/497846/ (Accessed March 28, 2022).
235. "Exploiting Earth-Moon Space: Chinese Ambitions after Space Station", Xinhua, March 08, 2016 at http://m.chinadaily.com.cn/en//2016-03/08/content_23775949.htm (Accessed March 28, 2022).
236. Ibid.

its lunar mission. This includes several landings on the South Pole of the Moon, towards establishing a research base. The Phase 4 will consist of the Chang'e 6, 7 and 8. Wu specified that:

> Chang'e-6 will attempt to retrieve one to two kilograms of samples from lunar pole regions and return them to earth. Chang'e-7 will land on the south pole of the moon, with the goal of looking for water ice and surveying the region's environment and landform. Chang'e-8 the phase finale will scout how to exploit the resources on the lunar south pole.

The south pole of the Moon is advantageous due to its peaks or eternal sunlight as well as presence of water-ice. Wu Weiren noted. "As the gravity on Moon is only one sixth of the Earth's, it is easier for probes to take off and the spacecraft can also solve the fuel supply issues on the moon."[237] China is building a lunar crew module for its human lunar missions and has been actively testing it, launched for an uncrewed test in May 2020.[238] A human landing on the Moon is now anticipated by 2030. As mentioned in the section on China's space goals, China is also developing a Mars human landing system, asteroid probes, and planetary defense.

Military Space Capacities

China views its space capacity as a critical component of its national power and directly linked to national security, one of the drivers for China's space program, identified in the earlier section of this paper.

---

237. Deng Xiaoci, "China to carry out Phase-4 lunar probe missions before 2030, build permanent station by 2035: project chief designer and political advisor Wu Weiren", *The Global Times*, March 08, 2022, https://www.globaltimes.cn/page/202203/1254281.shtml (Accessed March 28, 2022).
238. Andrew Jones, "China Reveals Details of Next Gen Crewed Spacecraft's 1st test Flight", Space.com, June 19, 2020, https://www.space.com/china-crew-spacecraft-test-flight-details.html (Accessed March 28, 2022).

The institution that directs China's space program is the State Administration on Science, Technology, and Industry for National Defense (SASTIND) which functions under the direction of the Ministry of Industry and Information Technology (MIIT).[239] As per its website, SASTIND is responsible for

> nuclear weapon, aerospace technology, aviation, armament, watercraft and electronic industries. It is established to strengthen military forces with additional personnel and more advanced equipment. Ensuring material supplies for the army is its top priority. Furthermore, it intends to contribute to the prosperity of the whole country by stimulating the manufacturing industry, gaining competitive edges through with superior production techniques. As the administrative and regulatory agency of science, technology and industry for national defense, SASTIND serves the needs of national defense, military forces, national economy, and military-related organizations. Meanwhile, it is also responsible for the coordination of communications and cooperation on the use of nuclear power and space activities with countries and international organizations.[240]

Under SASTIND is CNSA founded in 1993, and the other state funded space institutions like CAST, CALT, CASC. The PLA plays a critical role in developing China's space program including its human spaceflight program.

---

239. "Ministry of Industry and Information Technology", The State Council, The People's Republic of China, at http://english.gov.cn/state_council/2014/08/23/content_281474983035940.htm (Accessed March 28, 2022).
240. "State Administration for Science, Technology, and Industry for National Defense", The State Council, http://english.www.gov.cn/state_council/2014/10/06/content_281474992893468.htm (Accessed March 28, 2022).

President Xi Jinping has offered a clear policy guidance of 'Civil-Military Integration".[241] This last point is critical as Xi is clear about what he wants from such an integration policy: China's space activities, what Xi terms as "China's space dream" [now a part of his signature 'China Dream', and enshrined in the Communist Party of China (CPC) constitution][242] is to utilize China's space capabilities to create the economic resources to maintain China's leading position in the world. In short, China wants its space program to contribute to comprehensive national power and rejuvenation of the Chinese nation.[243] In fact, Xi who is the Chairman of the Central Military Commission (CMC) has frequently urged China's space sector to take up a leading role in his 'civil-military integration strategy'.[244] Xi also specifies that all space agencies (civilian and military) to include the 2015 established People's Liberation Army Strategic Support Force (PLASSF) owe complete loyalty and allegiance to the CPC.[245] In an article published in *Qiushi Journal*, the flagship journal of the CPC, for the 94th anniversary celebration of the founding of the PLA on

---

241. "Xi Calls for Deepened Civil Military Integration", *Xinhua*, March 12, 2018 at http://www.xinhuanet.com/english/2018-03/12/c_137034168.htm (Accessed March 28, 2022).
242. "Full Text of Resolution on Amendment to CPC Constitution", *China Daily*, October 24, 2017, http://www.chinadaily.com.cn/china/19thcpcnationalcongress/2017-10/24/content_33656521.htm (Accessed March 28, 2022). Also see Liu Mingfu, *The China Dream Great Power Thinking & Strategic Posture in the Post-American Era* (New York: CN Times, 2015).
243. "Full Text of White Paper on China's Space Activities in 2016", http://www.china.org.cn/government/whitepaper/node_7245058.htm (Accessed March 31, 2022). Also see Col. Qiao Liang and Col. Wang Xiangsui, *Unrestricted Warfare China's Master Plan to Destroy America* (Panama: Pan American Publishing Company, 2002).
244. "Xi Urges Efforts to Boost Integrated Military and Civilian Development", Xinhua, June 21, 2017, http://eng.chinamil.com.cn/view/2017-06/21/content_7646680.htm (Accessed March 28, 2022).
245. "Strive to Build a Strong, Modern Strategic Support Force: Xi", Xinhua, August 28, 2016, http://eng.chinamil.com.cn/view/2016-08/29/content_7231309.htm (Accessed March 28, 2022).

August 01, 2021, Xi stressed on the critical principle of PLA loyalty to the CPC as the core guide. Xi indicated that "the fundamental principle and system of the Party's absolute leadership over the armed forces are the political characteristics and rooted advantages of the PLA forces that completely distinguish them from all other old-style armed forces... realizing the centennial goal [CPC founding 1921-2021] is a major decision made by the CPC Central Committee and the CMC [Central Military Commission] in pursuing a stronger military. It is a major task relating to China's overall security and development, and a crucial step in the modernization of national defense and the armed forces."[246] China has demonstrated a capability by its SJ-21 satellite grabbing one of its defunct satellites in GEO and throw it 300 kms away into a graveyard orbit where satellites are unlikely to hit another functioning spacecraft. U.S. Space Command Commander, General James Dickinson stated in April 2021 that China's SJ-21 "could be used in a future system for grappling other satellites."[247] The dual use of satellites for something like debris clearing and grabbing an adversary satellite in times of conflict is a realistic scenario. On-Orbit Servicing, Assembly, and Manufacturing (OSAM) satellites, which SJ 21 is believed to be one, is such a dual capability.

---

246. Hu Yuwei and Xu Hailin, "Xi stresses CPC's leadership and achieving goals set for PLA centennial prior to Army Day", *The Global Times*, July 31, 2021, https://www.globaltimes.cn/page/202107/1230155.shtml (Accessed March 29, 2022). Also see "Xi Focus: Xi Calls on Young Officials to Strengthen Loyalty, Competency for Important Tasks", *Xinhua*, September 01, 2021, http://www.news.cn/english/2021-09/01/c_1310162324.htm (Accessed March 29, 2022).

247. Esteban Pardo, "Chinese 'space cleaner' spotted grabbing and throwing away old satellite", DW, February 09, 2022, https://www.dw.com/en/chinese-space-cleaner-spotted-grabbing-and-throwing-away-old-satellite/a-60658574 (Accessed March 30, 2022).

People's Liberation Army Strategic Support Force (PLASSF)

In 2015, under the direction of Xi Jinping, China institutionalized a separate space service with the establishment of the PLASSF. In his speech to the new service in August 2016, Xi emphasized that "the strategic support force bears a historic mission, it must strive to be the best in the world and be brave in innovation and exceeding others… the force should always stay on alert and maintain combat preparedness, map out a development strategy and a capacity building plan, build a new training system, and enhance its deterrence and warfighting capabilities".[248] He stressed on the need for innovation and creating new doctrinal concepts to steer the new service as it takes on new roles.

The PLASSF clubs together China's growing military space assets, into a single unit, aimed at dominance across the spectrum of air, space and cyber. The PLASSF is directed by Xi's Civil-Military Integration strategy. In 2017, the PLASSF signed a 'Cooperation Framework Agreement" with the China University of Science and Technology, Shanghai Jiaotong University, Xi'an Jiaotong University, Beijing Institute of Technology, Nanjing University, Harbin Institute of Technology and other six universities, as well as Aerospace Science and Technology Corporation, Aerospace Science and Industry Corporation, Electronic Technology Group Corporation.[249] These agreement is aimed at civil-military integration to attract talent, innovation, carry out special training

248. "Build a Strong, Modern Strategic Support Force: Xi", Xinhua, August 31, 2016, http://en.people.cn/n3/2016/0831/c90786-9108062.html (Accessed March 29, 2022).
249. "Strategic Support Force Cooperates with 9 Local Units to Train High End Talents of New Combat Forces", *Xinhua*, July 13, 2017, http://military.people.com.cn/n1/2017/0713/c1011-29402123.html (Accessed March 28, 2022).

based on technical expertise, and allow smooth transition between civilian agencies and the PLASSF.

The PLASSF fulfils four key goals for China. First, as per Xi's direction, the service is optimized to develop space innovation, develop a cislunar space situational awareness and presence capacity, and deny adversaries an advantage via their space system by engaging in jamming, dazzling, and RPOs.[250] Second, the PLASSF aims to develop futuristic doctrinal concepts for supporting China's growing space presence, to include its lunar and Mars missions. Towards this goal, partnership with China's private space sector as per Xi Jinping's direction since 2014.[251] Third, the PLASSF is focused on developing long term expertise on "big data applications, cloud computing, 3-D printing and nanomaterials. Always prepared for tomorrow's war". According to Yin Zhuo, Director of the PLA Navy's Expert Consultation Committee, "the PLA can maintain superiority in military operations taking place in space and involving the electromagnetic field and the internet…the force's responsibilities include target reconnaissance and tracking, global positioning operations, space assets management and defending against hostile activities in cyberspace and the electromagnetic field."[252] Fourth, PLASSF is being groomed as a war fighting force, both for terrestrial support for integrated PLA operations on land, sea, air, cyberspace, LEO and GEO space as well as cislunar space. Reports indicate that the PLASSF is training with ASAT missiles aimed at U.S. satellites. These ASAT includes the

---

250. M. Stoakes, "China's Space and Counterspace Capacities and Activities", U.S.-China Economic and Security Review Commission, March 30, 2020, https://www.uscc.gov/sites/default/files/2020-05/China_Space_and_Counterspace_Activities.pdf (Accessed March 29, 2022).
251. "PLA welcomes Launches of Private Rockets", *China Daily*, February 18, 2019, http://eng.chinamil.com.cn/view/2019-02/18/content_9429066.htm (Accessed March 29, 2022).
252. Zhao Lei, "Xi Calls New PLA Branch a Key Pillar", *People's Daily*, August 30, 2016, http://en.people.cn/n3/2016/0830/c90000-9107205.html (Accessed March 29, 2022).

variant of the HQ-19 surface to air missile, (utilized in ASAT tests in 2007, 2010) and the DN-2 (2013) and the DN-3 (2015, 2016, 2017). According to the National Space and Air Intelligence Center (NASIC) report, *Competing in Space*,

> China has military units that have begun training with anti-satellite missiles. Russia is probably also developing an anti-satellite missile. These missiles can destroy U.S. and allied space systems in low Earth orbit, making intelligence, surveillance, reconnaissance, and communications satellites vulnerable.[253]

Military Satellites and ASAT Weapons

China possesses nearly 50 military satellites. The functions of these military satellites can be divided into communication, early warning, meteorology, navigation, ocean surveillance, reconnaissance (optical and radar), Signal (SIGINT), Electronic (ELINT) and Communication Intelligence (COMINT). Among the communications satellites are FH 1A, 1B (ZX 22, 22A), FH 2A, 2B, 2C (ZX 1A, 1B, 1C), ST 1, 1B (ZX 20, 20A), ST 2A, 2B, 2C, 2D, 2E (ZX 2A, 2B, 2C, 2D, 2E). The *Feng Huo* (FH) series are data relay satellites that enables China to relay and pass data to ground receiving stations that are 'out of sight'. Built by CASC, this series provide military units with both C-band and Ultra-High Frequency (UHF) communications.[254] For early warning and detection, China possesses the Huoyan-1 series, of which the Tongxin Jishu Shiyan

---

253. National Air and Space Intelligence Center, "Competing in Space", December 2018, https://media.defense.gov/2019/Jan/16/2002080386/-1/-1/1/190115-F-NV711-0002.PDF, p.21 (Accessed on March 29, 2022).
254. "Feng Huo 1 (FH-1", https://www.globalsecurity.org/space/world/china/fh-1.htm (Accessed March 29, 2022).

Weixing (Communication Technology Test Satellite or TJSW) is viewed as co-linked for purposes of communication from GEO.[255]

An independent BeiDou (BDS) adds to China's military CnC as well. China can now independently guide missiles and bombs onto fixed targets without fear that the United States would turn off navigation services. BDS augments China's ability to guide its missiles very close to the target, after which a terminal seeker can provide active guidance for precise targeting. BeiDou augments independent military command and control by allowing precise knowledge of the location of one's own forces, and the ability to precisely target and provide navigation for military forces and strikes. This capability strengthens China's ability to coerce or compel others within its sphere of interest, such as on issues like the South China Sea, Taiwan or Hong Kong. An independent BDS coupled with 5G means real-time military command and control and devastatingly accurate automated weapons systems.[256] Consequently, the vital role that space plays for CnC, Positioning, Navigation and Timing (PNT), informationized warfare, space access and presence, space-based communications systems, and developing counter-space capabilities via-a-vis adversary is appreciated by China. For ocean surveillance and reconnaissance, China possesses the Yaogan, the Gaofen, the Tianhui, the Fanhui Shi Weixing (FSW), the Ludikancha Weixing (land surveying satellites or LKW), and the Ziyuan (ZY) series of military satellites. Yoagan series is built for Earth observation from LEO. These satellites are built by CAST and the Shanghai Academy of Space

255. Leo Bruce, "China Launches Chang Zheng 3B for Seventh TJSW Mission", NASA Spaceflight.com, August 24, 2021, https://www.nasaspaceflight.com/2021/08/tjsw-7-launch/ (Accessed March 29, 2022).
256. Namrata Goswami, "The Economic and Military Impact of China's BeiDou Navigation System".

Technology (SAST).[257] For SIGINT, COMINT and ELINT, China has the Yaogan series, the Tongxin Jishu Shiyan (TJS 1, 4 and 9), located in GEO and the Ji Shu Shiyan Weixing (JSSW), a series of experimental military satellites.[258]

In 2007, China tested its Anti-Satellite (ASAT) capability. Such ASAT technologies have been further refined in 2010, 2013 and 2014, enhancing their capabilities without generating space debris as did its 2007 test. There are open sources reporting that the SSF is training with ASAT missiles aimed at U.S. satellites. These ASAT includes the variant of the HQ-19 surface to air missile, (utilized in tests in 2007, 2010) and the DN-2 (2013) and the DN-3 (2015, 2016, 2017). "In 2019, the PRC described space as a "critical domain in international strategic competition" and stated the security of space provided strategic assurance to the country's national and social development."[259] Prioritizing the networked centric warfare approach, China is committed to developing its ASAT capacities in LEO and GEO.[260] Key bodies are the Space Systems Department in charge of PLA space operations, from launch, surveillance, missile tracking, and space warfare doctrine and concepts. Space capacities that China is developing for countering adversaries is kinetic kill,

---

257. Andrew Jones, "China Launches New Set of Yaogan Earth Reconnaissance satellites", Space.com, November 10, 2021, https://www.space.com/china-yaogan-35-reconnaissance-satellites-launch-november-2021 (Accessed March 29, 2022).
258. For details, please see Gunter Space Page, "Military Spacecraft China", https://space.skyrocket.de/directories/sat_mil_china.htm, https://space.skyrocket.de/directories/sat_mil_china.htm (Accessed March 29, 2022).
259. Annual Report to Congress, 2020. https://media.defense.gov/2020/Sep/01/2002488689/-1/-1/1/2020-DOD-CHINA-MILITARY-POWER-REPORT-FINAL.PDF (Accessed September 27, 2020). p.viii
260. Sandra Irwin, "Pentagon Report: China Amassing Arsenal of Anti Satellite Weapons", Spacenews, September 1, 2020, https://spacenews.com/pentagon-report-china-amassing-arsenal-of-anti-satellite-weapons/ (Accessed March 31, 2022).

direct ascent, co-orbital, electronic warfare (satellite jammers), ground-based lasers, space robotic capacities including a robotic arm, reconnaissance and surveillance, to create access for itself and deny access to adversaries in times of conflict. As stated earlier, the PLASSF is training with ASAT capabilities, and China's missiles systems create additive value. The SC-19 modeled on the DF-21C ballistic missiles is China's primary ASAT weapon, ranging between 2, 150kms-2, 500 kms.[261] Besides, China possesses the DF-26 Intermediate Range Ballistic Missile (IRBM), both a land and anti-ship missile, the DF-21A and DF 21E (nuclear tipped), and is predicted to possess 375 MRBM/IRBM missiles by 2025 up from 250 in 2020.[262] Most critically, about 75 of those missiles will be equipped with 'hypersonics' especially the DF-17.[263] China has also increased their nuclear warheads, from 290 (2019) to 320 (2020).[264] On July 27, 2021, the *Financial Times* reported that China had tested a nuclear-capable hypersonic glide vehicle that traversed LEO, circling Earth before striking close to its target at a speed of Mach 5.[265] China's Ministry of Foreign Affairs spokesperson, Zhao Lijian,

---

261. "SC 19 Anti-Ballistic Missile Interceptor," Global security.org, https://www.globalsecurity.org/space/world/china/sc-19-abm.htm (Accessed March 31, 2022).
262. "DF-26", Missile Threat, CSIS Missile Defense Project, https://missilethreat.csis.org/missile/dong-feng-26-df-26/ (Accessed March 31, 2022).
263. Hans M. Kristensen, "The (Other) Red Storm Risings: INDOPACOM China Military Projection," Federation of American Scientists, September 15, 2020 (Accessed March 28, 2022).
264. "Nuclear Weapon Modernization Continues but the outlook for arms control is bleak," SIPRI, June 15, 2020. https://www.sipri.org/media/press-release/2020/nuclear-weapon-modernization-continues-outlook-arms-control-bleak-new-sipri-yearbook-out-now (Accessed March 29, 2022). Also see Namrata Goswami, "Status of Existing and Emerging Asia Pacific Space Powers Capabilities", APLN, August 17, 2021, https://www.apln.network/projects/wmd-project/status-of-existing-and-emerging-asia-pacific-space-powers-capabilities (Accessed March 31, 2022).
265. Demetri Sevastopulo and Kathrin Hille, "China Tests New Space Capability with Hypersonic Missile", *The Financial Times*, October 16, 2021,

denied the report that it was a nuclear hypersonic glide vehicle, instead stating that it was the test of a reusable spacecraft.[266] Whereas U.S. Air Force Secretary, Frank Kendall stated that China's test could be similar to Fractional Orbital Bombardment System (FOBS) developed by the Soviet Union.[267] In July, 2021, China is believed to have launched a suborbital spaceplane from its Jiuquan Satellite Launch Centre that landed at an airport just over 800 kilometres away at Alxa League in Inner Mongolia Autonomous Region. CASC put out a press release in which it was stated that "on July 16, 2021, the suborbital reusable demonstration project carrier developed by the First Academy of China Aerospace Science and Technology Corporation Group Co. Ltd ignited and took off on time at the Jiuquan Satellite launch Center. After completing the flight according to the set procedure, it landed smoothly and horizontally in Alxa Youqi Airport, the first flight mission was a complete success".[268] Some trace this development as motivated by the U.S.'s X-37 robotic space plane, or Orbital Test Vehicle.

---

https://www.ft.com/content/ba0a3cde-719b-4040-93cb-a486e1f843fb (Accessed March 29, 2022).

266. "China denies Report of Hypersonic Missile Test, says Tested Space Vehicle", *Reuters*, October 18, 2021, https://www.reuters.com/world/china/china-disputes-report-hypersonic-missile-test-says-tested-space-vehicle-2021-10-18/ (Accessed March 29, 2022).

267. Shannon Bugos, "China tested Hypersonic Missile Capability: U.S. Says", *Arms Control Today*, November 2021, https://www.armscontrol.org/act/2021-11/news/china-tested-hypersonic-capability-us-says (Accessed March 29, 2022).

268. "我国亚轨道重复使用运载器飞行演示验证项目首飞取得圆满成功 (translated as 'the first flight of my country's suborbital reusable vehicle flight demonstration project was a complete success'), Press Release, CASC, July 16, 2021, https://mp.weixin.qq.com/s/ncJePW1znFexqkSbZ-0YRg (Accessed March 29, 2022).

Commercial Space

China has been developing its commercial space capacity under the direction of President Xi Jinping. As stated earlier in the paper, this was made of priority in 2014 when Document 60 was released by the State Council of the CPC to encourage China's private space sector.[269] Since then, there has been both larger investments in China's private space sector as well as a focus on civil-military integration.[270] Since 2016, China's private funding has seen an increase to about $3B annually.[271] Private space companies have achieved orbital launch and are invested in developing critical technologies like reusable space launch capacity. Amongst the known private space companies are Onespace, one of the first company to successfully launch to sub-orbit,[272] followed by attempts made by Linkspace and Landspace. In August 2019, Linkspace experimented with the first of China's reusable launch vehicle, when its rocket reached a height of 300 meters above ground and then landed back intact.[273] Beijing Interstellar Glory

---

269. State Council Information Office of the People's Republic of China, "Guiding Opinions of the State Council on Encouraging Social Investment in Investment and Financing Mechanisms in Key Innovation Areas Document 60." November 16, 2014. http://www.gov.cn/zhengce/content/2014-11/26/content_9260.htm (Accessed March 31, 2022).
270. Zhao Lei, "Civil-Military Integration will Deepen," *China Daily*, March 3, 3018 http://www.chinadaily.com.cn/a/201803/03/WS5a99d67ca3106e7dcc13f437.html (Accessed March 31, 2022).
271. Michael Sheetz, "China Increases Investment in Emerging Private Space Industry," CNBC, October 10, 2018, https://www.cnbc.com/2018/10/10/china-increases-investment-in-emerging-private-space-industry.html (Accessed March 31, 2022)
272. Reuters, "China Launches First Rocket Designed by a Private Company," May 17, 2018, , https://www.reuters.com/article/us-space-launch-china-onespace/china-launches-first-rocket-designed-by-a-private-company-idUSKCN1II0FK (Accessed March 31, 2022).
273. "China Successfully Launches Largest Reusable Rocket," *Global Times*, August 10, 2019, https://www.globaltimes.cn/content/1160951.shtml (Accessed March 31, 2022)

Space Technology Ltd or ispace launched its rocket *Hyperbola 1* into orbit in July 2019, marking the first such successful orbital launch by a Chinese private company.[274] China's ispace is aspiring to launch China's first reusable rocket into orbit, the *Hyperbola-2*.[275] In December 2021, Galactic Energy, a private space company built and launched its four stage *Ceres 1* rocket that placed five satellites. Also known as *Gushenxing-1,* the first *Ceres 1* rocket was successfully tested in November 2020 thereby becoming the second Chinese space company after ispace to launch a rocket to LEO.[276] Galaxy space, another Chinese space startup aims to launch 142 5G satellites into orbit. China is planning to establish a commercial spaceport as per its 14th Five Year plan as part of building a modern space infrastructure.[277]

Scenarios and Conclusions

In conclusion, based on the assessment of the drivers of China's space program, the goals and China's space capacity, it is evident that China aims to become a comprehensive great power in space

274. Reuters, "Chinese Rocket Startup Puts Satellites into Orbit for First Time," Reuters, July 25, 2019, https://www.reuters.com/article/us-china-space/chinese-rocket-startup-puts-satellites-into-orbit-for-first-time-idUSKCN1UK0KY (Accessed March 31, 2022).
275. "China's Private Reusable Rocket to be Launched in 2021," *Xinhua*, October 18, 2019, http://www.xinhuanet.com/english/2019-10/18/c_138482839.htm (Accessed March 31, 2022).
276. Justin Davenport, "China's Galactic Energy Launches Second Ceres-1 Rocket Successfully", NASA Spaceflight.com, December 07, 2021, https://www.nasaspaceflight.com/2021/12/chinas-galactic-energy-launches-second-ceres-1-rocket-successfully/ (Accessed March 29, 2022).
277. See "Outline of the 14th Five-Year Plan (2021-2025) for National Economic and Social Development and Vision 2035 of the People's Republic of China", August 9, 2021, https://www.fujian.gov.cn/english/news/202108/t20210809_5665713.htm#:~:text=The%20period%20covered%20by%20the,a%20moderately%20well%2Doff%20society. (Accessed March 31, 2022).

with civilian, military and commercial capabilities. These includes building its space infrastructure of lunar and Mars landings, invest in SBSP, asteroid mining and planetary defense. More importantly, China aims to build its own national satellite constellation prioritizing the critical importance of satellite internet. It has also developed policy to encourage its private space sector via funding and subsidized access to its PLA maintained launch sites. Most important of all, the vision for China's leadership in space comes directly from its President Xi Jinping, who time and again, have highlighted the need for China to innovate in technologies like space, Artificial Intelligence, and Quantum, to name a few. Xi's Civil-Military Integration strategy and the National Defense Law of 2021 are strategic moves to ensure that China's space investments build into China's comprehensive national power, a systems approach led and directed by the CPC.

China's investments in its military space capacity are aimed at ensuring that it offers China both compellence (inform adversary behavior to be in China's interest) and deterrence (deter adversary from assuming dominance in space or deter it from targeting its space systems). The development of its ASAT capabilities, its RPOs, blinding and dazzling, its satellites with robotic arms that can both clean debris or grab an adversary satellite in LEO and GEO, its independent navigation and satellite communication systems, and the establishment of PLASSF are clearly aimed at showcasing its military space capacities. China views space as a warfighting domain and President Xi Jinping has directed the PLASSF to develop new operational and doctrinal concepts to establish space superiority for purposes of power projection.

Chinese senior space leadership views space from an economic perspective especially highlighting the potential of the Moon and Mars and asteroid mining from a space utilization perspective. Given these vision, China's space programs are establishing programs that are geared towards taking advantage of 'first presence' concepts on the Moon and Mars, to establish the regulatory frameworks with China at the lead. Given China's past record of accomplishing its stated space goals, those who dismiss or ignore China's announced roadmaps for space, should consider the fact that achievements in space are directly connected to the CPC's legitimacy and are not taken lightly, given the high levels of political engagement with the space program. China's achievements in the field of science and technology are well documented.[278] While the U.S. is quick to dismiss China's achievements as based on allegations of theft of intellectual property, this perspective fails to take into account the fact that China has emerged as second only to the U.S. when it comes to investment in Research & Development (R&D).[279] In 2021, China increased its investments in R&D to 2.79 Trillion Yuan ($441 billion), an increase by 14 per cent over 2020 .[280] The case of Huaweii emerging as a leader in 5G despite the naysayers should give us pause. Ren Zhengfei, the founder of Huawei, employs nearly 80,000 of its staff for R&D alone, and has committed $15-20 billion on its R&D program.[281] This is augmented

---

278. "China's Quantum Satellite Clears Major Hurdle on way to Ultrasecure Communications", Nature, June 15, 2017 at https://www.nature.com/news/china-s-quantum-satellite-clears-major-hurdle-on-way-to-ultrasecure-communications-1.22142 (Accessed March 31, 2022). Louis Columbus, "How China is Dominating Artificial Intelligence", Forbes, December 16, 2018, https://www.forbes.com/sites/louiscolumbus/2018/12/16/how-china-is-dominating-artificial-intelligence/#550950de2b2f (Accessed March 31, 2022).
279. Matt Hourihan, "New Data Says U.S. R&D Has Topped 3% of GDP For the First Time Ever", American Association for the Advancement of Science, March 25, 2021, https://www.aaas.org/news/new-data-says-us-rd-has-topped-3-gdp-first-time-ever (Accessed March 31, 2022).
280. National Bureau of Statistics for China, "China's R&D Expenditure Reached 2.79 Trillion Yuan in 2021", http://www.stats.gov.cn/english/PressRelease/202201/t20220127_1827065.html (Accessed March 31, 2022).
281. Sijia Jiang, "China's Huawei to Raise Annual R&D Budget to at least $15 billion," *Reuters*, June 26, 2018, https://www.reuters.com/article/us-huawei-r-

by the rapid increase in engineering, science and tech graduates in China since 1999.[282] To be sure, those with engineering, science degrees are sought after and draw high start salaries. This is further supported by the Chinese government's 'Made in China 2025' policy.[283] CAS, which has about 120 other institutions under it, followed by Chinese Science and Technology universities and their privately owned industries are steering the course of space development in China.[284]

One of the scenarios that could develop as China and Russia establishes their research base [a permanent structure on the South Pole of the Moon, by 2036] is that they could deny that area for landing to another country by building a 'safety or non-interference zone'. Such Chinese behavior can be witnessed in regard to the East China Seas where they have created an Air Defense Identification Zone as well as their claims on the South China Sea Islands, where contrary to the fact that those islands are disputed, China has unilaterally decided that they have the right to build artificial permanent presence on these islands, and harden airfields to then install military hardware. Given the dominant position of China's military power in this region, there has been little counter-moves from other countries to include the U.S. in response to China's aggressive move on these islands, which stands uninterrupted today. What is there to stop China from behaving in a similar fashion on

---

d/chinas-huawei-to-raise-annual-rd-budget-to-at-least-15-billion-idUSKBN1KG169 (Accessed on April 08, 2019).

282. Katherine Stapleton, "Inside the World's Largest Higher Education Boom", *The Conversation*, April 10, 2017, https://theconversation.com/inside-the-worlds-largest-higher-education-boom-74789 (Accessed March 29, 2022).

283. U.S. Chamber of Commerce, "Made in China 2025: Global Ambitions Built on Local Protections", 2017, https://www.uschamber.com/sites/default/files/final_made_in_china_2025_report_full.pdf (Accessed March 31, 2022). Joachim Jan Thraen, *Mastering Innovation in China: Insights from History on China's Journey towards Innovation* (Wiesbaden, Germany: Gabler Verlag, 2016).

284. Ibid.

the Moon once they have the capacity to access its resources. International law is weak on this regard, and we have seen that nations can unilaterally interpret the Outer Space Treaty (OST) of 1967 to suit their national interest.

Finally, China under President Xi Jinping is a confident great power, and is focused on national rejuvenation and establishing the legitimacy of the CPC with an iron hold, both internally and internationally. Alignments along the BRI Spatial Information Corridor builds into China's image amongst the intended audience as a capable country with the resources to build the much-required space and information infrastructure, that these countries require. Consequently, the membership of the BRI is high, nearly 139 member nations. With space as part of the BRI, China's ability to influence regional and global events increase. To win without fighting is the strategy of a superior civilization, so states Liu Mingfu, the author of the book *China Dream*. It is evident that China is edging closer to space leadership, by recognizing the impact space has for national strategy and by prioritizing it as part of its infrastructure.

Japan and South Korea: Two Important Partners for the United States for Space Security in East Asia[285]

Sam Wilson

Introduction

Space partnerships play a critical role in U.S. national security space strategy. The Department of Defense's Defense Space Strategy, released in June 2020, emphasizes international space partnerships. One of the objectives of the strategy is to "integrate allies and partners into plans, operations, exercises, engagements, and intelligence activities."[286] The commitment to international defense partnerships is highlighted in other relevant strategic documents, including the 2018 National Defense Strategy and 2011 National Security Space Strategy.[287] Identifying and expanding areas for international cooperation is also a central theme in the 2020 National Space Policy. [288]

---

285. This paper was prepared for the 24th Air and Space Power International Conference held at Yonsei University.
286. Department of Defense, "Defense Space Strategy Summary," (June 2020).
287. Department of Defense, "Summary of the 2018 National Defense Strategy of the United States of America," (January 2018); Department of Defense and Director of National Intelligence, "National Security Space Strategy," (January 2011).
288. U.S. White House, "National Space Policy of the United States of America," (December 2020).

Space partnerships offer enormous advantages. Next year, the United States and Norway are planning to launch a Norwegian satellite that will be hosting U.S. national security payloads. This is projected to save the United States up to $900 million, and it will be the first time that the United States has ever put a national security payload on a foreign satellite.[289] In the next two years, the United States will be launching U.S. national security payloads on a Japanese satellite.[290] That will be the first time the United States has ever put a national security payload on a foreign satellite and foreign launcher. This saves money, it allows the allied and partner nations to leverage each other's unique capabilities and systems, and it also offers deterrence advantages – an adversary may think differently about attacking a system if that system is connected to multiple countries. [291]

Another important benefit of space partnerships is geography, and there is no region more important than East Asia for space security. China has been developing and testing counterspace weapons, and senior U.S. officials have said China is developing weapons for all orbits, including those of U.S. nuclear command and control satellites.[292] Moreover, the region has been an area of intense growth for space activity. In such a context, Japan and South Korea—important allies of the United States and established and growing space nations—represent crucial partners for space security.

---

289. Nathan Strout, "How 2 Space Norway satellites will help the Air Force in the Arctic." C4ISRNET (July 5, 2019).
290. Paul McLeary and Theresa Hitchens, "US, Japan To Ink Hosted Payload Pact to Monitor Sats," Breaking Defense, (August 5, 2019).
291. For more discussion on the importance of defense space partnerships for the United States, please see: Sam Wilson, Colleen Stover, and Steven R. Jordan Tomaszewsk, "Defense Space Partnerships: A Strategic Priority," The Aerospace Corporation, (September 2020).
292. For example, see Christopher Ford, "Whither Arms Control in Outer Space? Space Threats, Space Hypocrisy, and the Hope of Space Norms." Remarks at the Center for Strategic and International Studies. U.S. Department of State (April 6, 2020).

Long History of Japan's Shift Toward Space Security

Japan is one of the most mature space nations in the world and is becoming a major defense space nation. As of 2022, it has the third largest space budget and the fourth most satellites in orbit.[293] Japan has an array of dual use space assets. The country has its own reconnaissance satellites, is developing its own defense satellite communications system, and is pursuing a space situational awareness radar that will look out past geosynchronous Earth orbit (GEO). Japan's maturation as a space nation reflects its long history in space. In 1970, the nation launched "Ohsumi," its first satellite launched into orbit. The successful launch made Japan the fourth spacefaring country after the Soviet Union, United States, and France. Its maturation as a defense space power reflects a long shift toward focusing on defense space activity.

In the early stages of Japan's space program, the country sought to make a sharp distinction between its space and military activity. In 1969, the Diet, Japan's Parliament, passed the Peaceful Purposes Resolution, which excluded military space activity in response to the 1967 United Nations (UN) Outer Space Treaty. The treaty notes the "use of outer space for peaceful purposes," but Japan's interpretation was unique among the members of the treaty.[294] Most countries, including the United States, held that the agreement allowed military space activity if it was nonaggressive. Japan's

293. "Government space budgets driven by space exploration and militarization hit record $92 billion investment in 2021 despite covid, with $1 trillion forecast over the decade," Euroconsult, Government Space Programs, (January 2022); Union of Concerned Scientists Satellite Database. Union of Concerned Scientists. (Data updated as of May 1, 2022).
294. United Nations Office for Outer Space Affairs. 2222 (XXI). Treaty on Principles Governing the Activities of States in the Exploration and Use of Outer Space, including the Moon and Other Celestial Bodies.

1969	The Japanese Diet passes the Peaceful Purposes Resolution, adopting the position that its space activity was to be nonmilitary.
1984	Japan's military begins acquiring satellite imagery.
1998	North Korea successfully tests Taepodong missile.
2003	Japan launches first Information Gathering Satellite.
2006	North Korea fires seven ballistic missiles.
2007	China conducts anti-satellite weapon test.
2008	Japan passes Basic Space Law, officially allowing for military space activity and centralizing space policymaking.
2009	Japan issues the 2009 Basic Space Plan, which notes investments in reconnaissance satellites, secure satellite communications, and navigation satellites.
2010	Japan launches first of its navigation satellites, the Quasi-Zenith Satellite System.
2012	Japan revises law governing JAXA to allow the agency to collaborate with the Ministry of Defense.
2013	Japan releases the second Basic Space Plan, which reiterates the capabilities and improvements outlined in 2009 and adds requirements for space situational awareness.
2016	Japan launches first of its X-band defense communication satellites.
2018	Japan issues the 2018 National Defense Program Guidelines, which call for counterspace capabilities.

Source: Aerospace Corporation, Japan's Gradual Shift Toward Space Security | The Aerospace Corporation

position that its space activity was to be nonmilitary aligned with its post-World War II pacifist constitution.[295]

In the early stages of Japan's space program, the country sought to make a sharp distinction between its space and military activity. In 1969, the Diet, Japan's Parliament, passed the Peaceful Purposes Resolution, which excluded military space activity in response to the 1967 United Nations (UN) Outer Space Treaty. The treaty notes the "use of outer space for peaceful purposes," but Japan's interpretation was unique among the members of the treaty.[296] Most countries, including the United States, held that the agreement allowed military space activity if it was nonaggressive. Japan's position that its space activity was to be nonmilitary aligned with its post-World War II pacifist constitution.[297]

Although the Peaceful Purposes Resolution remained intact for nearly 40 years, Japanese leadership across administrations made a series of decisions beginning in the 1980s that softened the lines between their space and military activity.

Generalization Theory

The Japanese government started buying imagery from the U.S. LANDSAT satellites in 1984 and from French SPOT satellites in 1987 for multiple agencies, including the Japanese Defense Agency, the predecessor to the Ministry of Defense.[298] The Yasuhiro

---

295. Robert S. Wilson, "Japan's Gradual Shift Toward Space Security," The Aerospace Corporation, pages 2-4. (May 2020).
296. United Nations Office for Outer Space Affairs. 2222 (XXI). Treaty on Principles Governing the Activities of States in the Exploration and Use of Outer Space, including the Moon and Other Celestial Bodies.
297. Robert S. Wilson, "Japan's Gradual Shift Toward Space Security," The Aerospace Corporation, pages 2-4. (May 2020).
298. Saadia M. Pekkanen and Paul Kallender-Umezu, "In Defense of Japan: From the Market to the Military in

Nakasone administration maintained this was consistent with the 1969 resolution.[299] As explained by Japanese space scholar Setsuko Aoki, "In 1985, a unified governmental view was issued that interpreted the 'nonmilitary' principle as permitting the use of satellites so long as these satellites were widely used in everyday civil life." The Japanese military could use satellite technologies that were readily available on the commercial or civilian side. This position, called the "generalization theory," was an important step toward the country's embrace of defense space activity.[300]

Development of Reconnaissance Satellites

In 1998, the Keizo Obuchi administration made the decision to develop government-owned and operated reconnaissance satellites called Information Gathering Satellites.[301] Although the satellites also had civilian purposes—such as monitoring natural disasters and weather patterns—this was the first time in which Japan acquired satellites that would have a direct defense application. The first two satellites were launched in March 2003, and the newly formed Cabinet Satellite Intelligence Center would oversee the reconnaissance satellites. Because the new center was not part of the

---

Space Policy," Stanford University Press, Stanford, California: 2010, pages 143-144.

299. Setsuko Aoki, "Introduction to the Japanese Basic Space Law of 2008 / Uberblick uber das Japanische Weltraumbasisgesetz 2008 / Introduction a la Loi (de Base) Spatiale 2008 Japonaise," Zeitschrift fur Luftund Weltraumrecht - German Journal of Air and Space Law 57, no. 4 (2008): 585-589.

300. Manuel Manriquez, "Japan's Space law Revision: the Next Step Toward Re-Militarization?" Nuclear Threat Initiative, January 1, 2008; and Robert S. Wilson, "Japan's Gradual Shift Toward Space Security," The Aerospace Corporation, pages 4-6. (May 2020).

301. Andrew L. Oros, "Explaining Japan's Tortured Course to Surveillance Satellites," Review of Policy Research 24(1):29-48, January 2007.

defense agency but part of the Cabinet Secretariat, the oversight stayed within the parameters of the 1969 resolution.[302]

Although there are competing explanations for the development of reconnaissance satellites, North Korean missile tests in the 1990s stand out for having an important effect among the Japanese. A 1998 test particularly made it clear that North Korea had the capability to hit targets almost anywhere in Japan. These tests contributed not just to Japan's development of its reconnaissance satellites but also its ballistic missile defenses, which the country decided to adopt in 2003.[303]

Passage of Basic Space Law

The Basic Space Law in 2008 is often regarded as a pivotal moment in the country's push toward space security. The law reversed the 1969 Peaceful Purposes Resolution, allowing for military space capabilities if they were "nonaggressive." It also acknowledged the importance of using space for national security purposes. According to an unofficial translation, the law asserts: "The State shall take necessary measures to promote space development and use to ensure international peace and security as well as to contribute to the national security of Japan."[304]

---

302. Robert S. Wilson, "Japan's Gradual Shift Toward Space Security," The Aerospace Corporation, pages 6-7. (May 2020).
303. Robert S. Wilson, "Japan's Gradual Shift Toward Space Security," The Aerospace Corporation, page 8 (May 2020).
304. A useful briefing on the law was provided by the Japanese government to the United Nations Committee of the Peaceful Use of Outer Space, Legal Subcommittee in 2009.

Counterspace and Missile Warning

In recent years, Japan has expressed interest in capabilities that would almost exclusively be used for defense purposes. Along these lines, Japan's guidance reflects interest in developing counterspace systems and missile warning satellites.

The 2018 National Defense Program Guidelines was the first official Japanese document to call out a need for counterspace systems. Specifically, the guidelines say that the Self Defense Forces will work to improve defense capabilities, including the "capability to disrupt opponent's command, control, communications, and information."[305] Experts have said that National Defense Program Guidelines referred to jamming technologies that the Ministry of Defense is considering acquiring.[306]

In 2020, the Diet approved the newest version of the country's national space policy, which was last updated six years prior. The new policy says the country will study small-satellite constellations with infrared sensors for missile warning. In 2021, Japan released its implementation plan of the national space policy, which also discussed missile warning. Japan developing its own early warning satellites would be significant. Currently, only the United States and Russia operate early warning satellite systems. Among other things, it would signal Japan's maturation as a defense space power.[307]

---

305. Japan Ministry of Defense, "National Defense Program Guidelines for FY 2019 and beyond," December 18, 2018.
306. Robert S. Wilson, "Japan's Gradual Shift Toward Space Security," The Aerospace Corporation, page 13 (May 2020).
307. Robert Sam Wilson, "OPINION: Japan's studying missile warning fits within long-term trends," Kyodo News (August 28, 2020).

South Korea's Rapid Acceleration in Space

South Korea is also emerging as an important defense space nation, albeit in a much more accelerated fashion and with a later start than Japan. In 2013, South Korea launched its first space launch vehicle that reached Earth orbit.[308] In 2022, for the first time, the country launched its own satellite from its own domestically built rocket, putting it in a small exclusive group of countries that can independently launch into orbit.[309] Despite only having a small number of active satellites, the nation is planning to increase that number dramatically.[310]

Among the 10 biggest spenders in space, South Korea its own military communications satellites and earth observation satellites and has ambitious plans for its space program. Its reported space budget for 2022 is $619 million, a 19 percent increase from the prior year.[311] This budget will support launch vehicle development, Earth observation satellites, communication satellites, a new satellite operations center, and the Korean Positioning System (KPS). KPS would be South Korea's space-based positioning, navigation, and timing (PNT) system. The successful fielding of those satellites would make South Korea one of six nations that has its own PNT satellites, along with the United States, Russia, China, India, and Japan. The country is reportedly pursuing eight KPS satellites; seven

308. Choe Sang-Hun, "On 3rd Try, South Korea Launches Satellite Into Orbit," The New York Times, (January 30, 2013).
309. China, India, Japan, Russia, South Korea, the United States, and the European Space Agency have independent
space launch capabilities to reach medium Earth orbit and GEO. "Competing in Space," National Air & Space Intelligence Center,
December 2018
310. Union of Concerned Scientists Satellite Database. Union of Concerned Scientists. (Data updated as of May 1, 2022).
311. Park Si-soo, "South Korea's double-digit space budget boost," SpaceNews (April 21, 2022).

would be necessary to deploy an independent regional system.[312] PNT satellites, similar to communications and earth observation satellites, are inherently dual use.

Much of the focus in South Korea is on commercial policy. In August 2022, in the release of the budget, the Ministry of Science and ICT notes the "acceleration toward the private led space economy."[313] In July 2022, the ministry announced five key action plans for the administration, which included transferring space technologies from the government to the private sector. Specifically, it said: "The space industry, originally led by the government, will nurture companies with many responsibilities (manufacturing + launch operation) and promote the development of space economy, by transferring technology and designating clusters (let the private sector utilize infrastructure)."[314] In June 2022, the South Korean National Assembly passed an amendment to the country's Space Development Promotion Act that largely focused on commercial activity. The amendment calls for the "establishment of space infrastructure for space development and provision of open access to the private sector…introduction of a new contract format to ensure companies' profitability…designation of new space technologies and acceleration of technology transfer…" and "promotion of talent nurturing and entrepreneurship."[315]

This focus on commercial space ventures had already some success. South Korean company Hanwha Systems, for example, announced

---

312. Park Si-soo, "South Korea's double-digit space budget boost," SpaceNews (April 21, 2022).
313. South Korea Ministry of Science and Technology, "MSIT plans to focus on national digital innovation and super gap technology securement," (August 31, 2022).
314. South Korea Ministry of Science and Technology, "MSIT to Announce Five Key Actions Plans of the New Administration," (July 15, 2022).
315. South Korea Ministry of Science and Technology, "The Amendment to the Space Development Promotion Act to pass the National Assembly," (June 7, 2022).

plans to deploy a constellation of 2,000 communication satellites in low Earth orbit by 2030.[316] (For context, as of May 2022, no country outside of the United States has more than 2,000 satellites in orbit.) Hanwha is involved in a separate project to build small surveillance satellites, which would have civilian and military applications, including monitoring North Korea.[317] Through a technology transfer from the state-funded Korea Advanced Institute of Science and Technology, the firm also plans to develop laser inter-satellite links. The Korea Advanced Institute of Science and Technology also has a partnership with Korea Aerospace Industries, a private company, to build a space research center. In 2021, Korea Advanced Institute announced plans to invest $880 million over the next five years to expand its space business, including satellite production, and hopes to develop a ground station and satellite imagery analysis services in the next few years.[318]

Economic indicators suggest South Korea's GDP will continue to fare relatively well, which could help its industry's acceleration in space. South Korea is not only one of the biggest economies in the world, but its economy is growing at a high rate relative to other countries with high GDP. Among the countries with the top 15 highest GDPs, only India has a higher rate of growth in GDP than South Korea.[319]

---

316. Park Si-soo, "Hanwha Systems to launch 2,000 LEO communications satellites by 2030," SpaceNews (March 30, 2021).
317. Brian Kim, "With restrictions lifted, South Korea launches $13B space power scheme," Defense News (September 6, 2021).
318. Brian Kim, "With restrictions lifted, South Korea launches $13B space power scheme," Defense News (September 6, 2021).
319. World Bank data. "South Korea."

Opportunities for Defense Space Partnerships for the United States

Although the United States has collaborated with Japan and South Korea, there is room for growth.

Japan and the United States

Japan and the United States have a robust defense space partnership. The two countries have had a space situational awareness agreement since 2013.[320] In recent years, Japanese military forces have taken U.S. Space Force space operations courses, including courses in space situational awareness and orbital mechanics, and have participated in space security exercises and wargames, such as the situational awareness exercise called Global Sentinel and the space wargames at Schriever Air Force Base.[321] Further, as mentioned, the United States is placing U.S. national security payloads on Japanese satellites (Japan's PNT satellites) in the next two years.

Despite these successes, there are some areas where the partnership could expand.

(1) Space situational awareness represents a promising area for further collaboration. Japan is developing an ambitious deep-space radar system and command and control center. Japan's fiscal year 2022 Ministry of Defense budget dedicates about $226 million to this effort, almost 40 percent of the ministry's defense space budget.[322] Given the counterspace threats in GEO, Japan's system

---

320. "Joint Statement – The First Meeting of the Japan U.S. Comprehensive Dialogue on Space," Japan Ministry of Foreign Affairs, (March 11, 2013).
321. Robert S. Wilson, "Japan's Gradual Shift Toward Space Security," The Aerospace Corporation, page 11 (May 2020).
322. Japan Ministry of Defense, *Defense Programs and Budget of Japan - Overview of FY2022 Budget*, p,4 (March 2022).

could be extremely advantageous for the United States. The United States has had challenges ingesting space situational awareness data from allies and partners, but it is taking some steps along these lines to mitigate these challenges.[323]

(2) Missile warning is another area in which the United States and Japan could partner more extensively. Japan's heightened interest in the mission area comes at a time in which the United States is beginning to develop and transition to a new missile warning and tracking architecture. The U.S. Department of Defense's fiscal year 2023 budget request includes funding for missile warning and tracking programs at GEO, Polar, medium Earth orbit, and low Earth orbit.[324] Such a diversified architecture could create opportunities for collaboration or joint development.[325]

South Korea and the United States

South Korea and the United States have also cooperated on space. Although this has not yet extended to developing capabilities jointly, the amount of engagement between the countries on space issues has increased over the last few years.

Notably, in April 2022, the South Korea's Defense Ministry and the U.S. Department of Defense –at the 18th session of the Space Cooperation Working Group—agreed to carry out research for a joint official document on space policy. "The representatives agreed

323. Lauren Hale, Justin C. Last, Jordan L. De Namur-Paul, and Richard Barber, "Partnering Not Bossing: Leveraging International Capabilities for Space Domain Awareness," The Aerospace Corporation (August 2021).
324. Robert S. Wilson, "Fiscal Year 2023 Space Force Budget Analysis: Missile Warning and Tracking Looms Large," The Aerospace Corporation (September 20, 2022).
325. Sam Wilson, "Opportunities for U.S.-Japan and U.S.-ROK Space Cooperation: Talk for Hudson Center and Nonproliferation Policy Education Center," NPEC and Hudson Institute (June 2022).

on the 'South Korea-US joint research on space policy' for the first time to develop space policy at the South Korea-US alliance-level," South Korea's Defense Ministry said in a Korean-language statement.[326] The representatives included Cho Yong-geun, the South Korean director general of the North Korea Policy Bureau at the Defense Ministry, and U.S. Deputy Assistant Secretary of Defense for Space and Missile Defense John Hill.[327]

PNT is an area that has generated some momentum for further collaboration. In August 2022, the Minister of Science and ICT met with Chirag Parikh, the Executive Secretary of the U.S. National Space Council. They reportedly discussed private space ventures and discussed collaboration for KPS.[328] In 2021, the two countries signed a joint statement on cooperation for global navigation satellite systems, in which they "jointly decided to support Korea's development of its own satellite navigation system, KPS, and enhance its compatibility and interoperability with the Global Positioning System (GPS)."[329] A regional independent PNT capability over South Korea could be particularly valuable to the United States as an alternative to GPS.

The countries are exploring and pursuing collaboration in other areas as well. In August 2021, the two countries signed a memorandum of understanding on forming a joint space policy consultative body at Peterson Air Force Base in Colorado Springs, Colorado. Under the agreement, the two sides have reportedly "run a joint consultative body on space policy, shared information on

---

326. Ji Da-gyum, "S. Korea, US agree on first joint research to develop defense space strategy," The Korea Herald (April 26, 2022).
327. Ji Da-gyum, "S. Korea, US agree on first joint research to develop defense space strategy," The Korea Herald (April 26, 2022).
328. South Korea Ministry of Science and Technology, "MSIT to buyild strong foundation for 'Korea-U.S. tech alliance,'" (August 2, 2022).
329. South Korea Ministry of Science and Technology, "Signing Ceremony for Korea-U.S. Joint Statement on Civil Global Navigation Satellite Systems Cooperation,'" (May 27, 2021).

space surveillance and worked together to enhance joint space operations capabilities such as missile defense."[330] In April 2022, the two countries announced an agreement that included cooperation on space situational awareness for military purposes—to "share intelligence about outer space, nurture space experts through training and exercises, and enhance interoperability for combined space operations."[331]

Implications of More Extensive Partnerships

More extensive defense space partnerships with Japan and South Korea will offer important advantages for the United States as well as for Japan and South Korea. As discussed, integrated capabilities like hosted payloads can generate savings and deterrence advantages. The advantages of these partnerships could also benefit broader geopolitical relationships. With Japan, the United States recently agreed to extend its security umbrella to outer space, ensuring that the United States would seek to protect Japanese satellites.[332] The partnership with Japan may serve as forerunner for what may be possible with South Korea. For example, hosting U.S. national security payloads on Japan navigation satellites may create a precedent in which the United States could host GPS payloads on South Korea's proposed navigation satellites.

In addition to bilateral efforts, there are also opportunities for trilateral space collaboration. In August 2022, Japan, South Korea, and the United States participated in a missile warning and ballistic missile and tracking exercise, which followed a trilateral ministerial

---

330. Park Si-soo, "US, South Korea agree to enhance security cooperation in outer space," SpaceNews (August 30, 2021).
331. Park Si-soo, "U.S., South Korea agree to cooperate on space situational awareness for military purposes," SpaceNews (April 26, 2022)
332. Miki Reiko, "Japan, U.S. extend security treaty into space to protect satellites," Nikkei Asia, (January 12, 2023).

meeting in Singapore.[333] Even more recently, in September 2022, Japanese Prime Minister Fumio Kishida met South Korean President Yoon Suk-yeol for a short meeting at the UN General Assembly.[334] As the United States treats defense space partnerships as more of a priority, bilateral efforts could extend to multi-lateral partnerships in the region, such as with Australia and New Zealand.

Conclusion

U.S. defense space partnership with Japan and South Korea will continue to grow more valuable in the years to come. The importance of the domain and the region behooves the United States to push for more defense space collaboration with these crucial partners. Japan and South Korea have been extremely important allies for the United States; in fact, both rank among the top three countries in the world for the number of military personnel the United States stations there. Although the United States, Japan, and South Korea have made significant strides for collaborating on space systems, there is ample room for growth.

---

333. U.S. Department of Defense, "U.S., Republic of Korea, and Japan Participate in Missile Defense Exercise in Hawaii (August 15, 2022).

334. Timothy W. Martin and Peter Landers, "Japan and South Korea Leaders Meet One-on-One for the First Time Since 2019," The Wall Street Journal (September 22, 2022).

PART III

A Synergistic Approach: Diplomatic and Unilateral Measures to Counter Chinese Anti-Satellite Spacecraft, Lasers, and Jammers

Brian G. Chow and Brandon W. Kelley[335]

Abstract

China's destructive test of a direct-ascent anti-satellite (DA-ASAT) missile in 2007 sent an unmistakable signal that the era of space-as-sanctuary was over. Since that time, however, the United States has failed to ready adequate and timely counters to China and Russia's expanding portfolios of next-generation ASAT weapons. While significant progress has been made, it remains halting, and primarily focused on the threats of yesterday and today rather than the near future. One key cause of this failure is parochial divisions between "hawkish" and "dovish" approaches, when in reality the new threat vectors now emerging demand a synergistic approach synthesizing elements of both. A second, related problem is the tendency to paint with a broad brush, rather than attending carefully to the unique characteristics of individual emerging threat vectors. While the United States has managed to muddle through thus far,

335. The authors thank the Nonproliferation Policy Education Center for commissioning and financially supporting this research paper. They also appreciated the opportunity to participate in the Center's China Space Wargame, as the exchanges of ideas with a diverse group of participants helped the authors better see others' concerns about space threats and others' preference on the types of solutions.

there is reason to believe that absent changes in course this may not remain the case much longer. Therefore, Nonproliferation Policy Education Center (NPEC) commissioned this report to evaluate three specific near-term space threats over the rest of this decade—namely, Chinese rendezvous spacecraft, ground-based lasers, and earth-based jammers—and illustrate the urgent need for this synergistic approach. Each threat is assessed in a nuanced manner on a case-by-case to identify what tailored combination of unilateral and diplomatic measures are necessary to counter it.

I. Introduction

China's successful direct-ascent anti-satellite (DA-ASAT) missile test in 2007 was a wakeup call for many regarding the vulnerability of United States space assets. Since then, substantial attention and financial resources have been dedicated to countering the threat. On the other hand, China and Russia have strong incentives to improve their counterspace capabilities to asymmetrically counter the U.S.'s superior military and economic power, as well as to hedge against the possibility of space-based deployment of missile defense and ground-targeted strike systems. However, while the U.S. has remained primarily focused on the DA-ASAT threat vector, potential adversaries are not only continuing to improve their existing ASAT weapons but developing and maturing new ones. Over the last seven years, many studies[336] by the two authors, singly or jointly, have shown signs that the West's preparations are highly unlikely to be adequate, or be ready in time, to counter these next-generation ASAT threats. The potential consequences could be catastrophic: acute use-or-lose pressures and political incentives could drive miscalculation and rapid escalation by either side,

---

336. A list of these studies and their links is available upon request to brianchow.sp@gmail.com or bwk9@georgetown.edu.

possibly resulting in large-scale conventional or even nuclear conflict; or new threat vectors might induce the U.S. to sacrifice vital strategic interests in order to avoid unacceptable losses of life and treasure, providing adversaries an irresistible means of achieving their strategic goals without firing a shot.

Development of these new ASAT threat vectors[337] is ushering in a new era of space security, characterized by exponentially greater complexity, instability, and risk. Yet, the conversation around space security has continued to be characterized by a stark divide between so-called "hawks," bent on unrestrained freedom-of-action and eschewing diplomatic engagement, and "doves," with a kneejerk aversion to unilateral action and *realpolitik*. This divide poses a serious threat to the national security of the United States as well as to the continuance of stability, peace, and prosperity in space and on Earth. This dichotomy was always a false one, but in the emerging era is set to make that painfully clear. A synergistic approach to address one specific threat at a time can be more efficiently than dove- or hawk-centric approaches to capture the strengths of both dovish and hawkish ideas and measures.

---

337. For a summary of these capabilities, refer to: Brian Weeden and Victoria Samson, eds., *Global Counterspace Capabilities: An Open Source Assessment*, (Washington, DC: Secure World Foundation (SWF), April 2022), https://swfound.org/media/207350/swf_global_counterspace_capabilities_2022_rev2.pdf; Todd Harrison et al., *Space Threat Assessment 2022*, Aerospace Security Project (Washington, DC: Center for Strategic & International Studies (CSIS), April 2022), https://csis-website-prod.s3.amazonaws.com/s3fs-public/publication/220404_Harrison_SpaceThreatAssessment2022.pdf?K4A9o_D9NmYG2Gv98PxNigLxS4oYpHRa; "Challenges to Security in Space," (Washington, DC: U.S. Defense Intelligence Agency (DIA), January 2019), https://apps.dtic.mil/sti/pdfs/AD1082341.pdf ; "2020 China Military Power Report," (Washington, DC: U.S. Department of Defense (DOD), September 2020), https://media.defense.gov/2020/Sep/01/2002488689/-1/-1/1/2020-DOD-CHINA-MILITARY-POWER-REPORT-FINAL.PDF.

So far, there have been some tentatively promising signs of groundwork of this synergistic approach being laid. One example may be found in Vice President Harris's April 2022 declaration banning debris-producing DA-ASAT tests—a measure which, though strictly diplomatic, was undertaken unilaterally as a first step, and was carefully-calibrated to address a specific threat *without* unduly tying the U.S.'s hands. Another is the elaboration of five tenets of responsible behavior set forth by Secretary Lloyd Austin in July 2021 to guide DoD space operations—operating with due regard, limiting long-lived debris, avoiding harmful interference, maintaining safe separation and trajectory, and enhancing space stability—[338] and efforts to advance and operationalize these tenets, both within the U.S. Government and in international relations. Nevertheless, to-date these measures have remained largely abstract, and little in the current dialogue focuses on specific emerging counterspace threats and *outlines concrete solutions*.

This report, commissioned by NPEC, seeks to contribute to a pivot toward the latter. This paper contends that the U.S. should adopt a synergistic approach to emerging ASAT threats characterized by three features. First, analysts should carefully evaluate the characteristics of each individual threat vector. With the threat landscape poised to change in radical, sudden, and unpredictable ways, "one-size-fits-all" approaches will not be sufficient. It is imperative to carefully consider what unique threat characteristics might render those broad defenses moot or require additional, tailored solutions. Second, assessments and recommendations must be *nuanced* and bipartisan, recognizing that unilateral measures—defined here as those which a country can carry out on its own without the consent of any other country—are not zero-sum with

338. Lloyd Austin, "SUBJECT: Tenets of Responsible Behavior in Space," Memorandum for the DOD, Office of the Secretary of Defense (OSD), July 7, 2021, https://media.defense.gov/2021/Jul/23/2002809598/-1/-1/0/TENETS-OF-RESPONSIBLE-BEHAVIOR-IN-SPACE.PDF.

bilateral or multilateral diplomatic efforts. Rather, they are mutually-reinforcing and interdependent; neither can be effective without the other. Finally, it must be primarily solution-oriented. Because identification of a threat is merely the first step in a long chain of traditionally time-consuming moves—e.g., selecting a solution, developing an accompanied concept of operations, carrying out the acquisition, and deploying the necessary defenses and deterrent measures—the U.S. cannot afford any significant lag in predicting or identifying threats, or between threat identification and solutioning.

As resource limitations preclude any single paper from providing an exhaustive assessment of all current and future threats, this report elects to focus narrowly on putting this methodology into practice for three threat vectors, against which our preparations are behind. Specifically, this paper will assess the threat posed during the 2020s by R-spacecraft, ground-based lasers, and earth-based jammers. For similar reasons, this paper will focus primarily on the China context, to the exclusion of other actors.

This does not imply that other ASAT threats—e.g., Russian co-orbital systems, Russian and Chinese offense cyber capabilities, their direct-ascent systems, Chinese cis-lunar threat vectors, and multiple others—are unimportant. However, these three threats, while they differ substantially in both type and severity of impact, each in their own way requires greater attention and more directed preparation to counter than they are currently receiving.

This report proceeds as follows: Section II examines the characteristics and potential implications of these three ASAT threat vectors, with particular emphasis on the role they are likely to play in China's efforts to counterbalance U.S. military and economic superiority. If the U.S. remains on its present course, it may fail in the latter half of the decade to preserve peace and stability by adequately deterring or countering the threat or use of these systems.

Section III outlines diplomatic and unilateral measures that can counter these three space threats. Finally, Section IV summarizes this paper's primary recommendations.

II. China's Anti-Satellite Spacecraft, Lasers, and Jammers Over the Rest of This Decade

This section examines the following three counterspace threat vectors: rendezvous spacecraft (R-spacecraft), ground-based lasers, and earth-based jammers. Each of these threats is defined, briefly explained, and their potential impacts assessed, with particular emphasis on their possible implications for Sino-U.S. military and economic competition in the 2020s.

A. Rendezvous Spacecraft

(i) Threat Profile

This report opts for the term "rendezvous spacecraft" to refer generally to satellites capable of conducting sophisticated on-orbit maneuvers for close-approach to other satellites, commonly known as "rendezvous and proximity operations" (RPOs), and especially to those satellites which are capable of direct manipulation or grappling of another satellite. Development and deployment of such RPO-capable spacecraft is an inevitability[339] and cannot realistically be prevented. Moreover, even if the development and deployment of such spacecraft *could* somehow be delayed or prohibited, doing so would severely hinder development of the space economy. Many countries—including the United States, China, Russia, European Union, and Japan, among others—have either already deployed such

---

339. Brian Chow, "Space Arms Control: A Hybrid Approach," *Strategic Studies Quarterly*, vol. 12, no. 2 (2018): 128, http://www.jstor.org/stable/26430818.

spacecraft or will in the next few years, for a variety of missions including transporting, *in-situ* repair, refueling, and servicing of satellites on-orbit and active debris removal (ADR).[340] These capabilities are crucial for facilitating sustainable and exponential growth in on-orbit activities.[341]

However, while these technologies and techniques hold immense promise for peaceful purposes, they are inherently dual-use: if a satellite is capable of docking with a friendly satellite to "tow" it into a new orbit for end-of-life disposal or to serve a new region on Earth, it can also grapple an adversary's satellite and move it to a "useless" orbit where it can no longer provide its intended service. If a spacecraft can finely manipulate space objects for servicing and repair, it can bend or disconnect an antenna or a solar panel to degrade or disable a satellite's function.

Many commentators have noted that taking spacecraft designed to perform highly complex and sensitive *civil and commercial* functions and employing them as ASATs is unlikely to be cost-effective.[342] Obviously, using a design that is optimized for all peaceful uses to perform ASAT purposes would be far less cost-effective than one optimally designed for ASAT use only. However, it is safe to assume that China has long recognized this situation and, thus, would not be interested in a single design optimized for all peaceful uses. There is also clear evidence, outlined below, that

---

340. Morgan Stanley, "Space: Investing in the Final Frontier," 24 July 2020, https://www.morganstanley.com/ideas/investing-in-space .

341. Joshua P. Davis, John P. Mayberry, and Jay P. Penn, "On-Orbit Servicing: Inspection, Repair, Refuel, Upgrade, and Assembly of Satellites in Space," The Aerospace Corporation, April 2019, https://aerospace.org/sites/default/files/2019-05/Davis-Mayberry-Penn_OOS_04242019.pdf.

342. James Alver, Andrew Garza, and Christopher May, "An Analysis of the Potential Misuse of Active Debris Removal, On-Orbit Servicing, and Rendezvous & Proximity Operations Technologies," Secure World Foundation, July 2019, https://swfound.org/media/206800/misuse_commercial_adr_oos_jul2019.pdf.

China has a dedicated ASAT R-spacecraft program. However, China is likely to argue that dual-use R-spacecraft of different number, size, cost, and capabilities are required to optimally serve the many different on-orbit satellite services including space debris removal. Even a given satellite service type could require spacecraft of different attributes to provide in different situations of use, such as serving small or large satellites. Such a complicated portfolio of dual-use R-spacecraft of different attributes will ensure that we cannot reliably distinguish development, testing, or deployment of ASAT-capable R-spacecraft from peaceful ones. Thus, the primary threat is that countries will able to freely develop, test, build, and ultimately deploy a large number of R-spacecraft designed and optimized specifically for counterspace use. These spacecraft may then be positioned in close proximity to our satellites, and, at the time of their choosing, these close-in spacecraft could execute a massive bolt-from-the-blue attack. This is a capability that China does not currently have, and could well be a game-changer. It is also likely to produce serious risks of miscalculation and inadvertent escalation if sufficient frameworks for managing the situation are not established and publicly articulated well in advance. It is worth noting that space-based ASAT weapons themselves are nothing new: the Soviet Union began development of its first co-orbital ASAT system, the *Istrebitel Sputnikov (IS)*, in 1961.[343] However, a number of current and near-future developments substantially change the picture. These include increasingly sophisticated rendezvous

---

343. Anatoly Zak, "IS Anti-satellite System," last modified July 13, 2017 (Russian Space Web), http://www.russianspaceweb.com/is.html. The full set of Russian co-orbital programs including this one are covered at length by the "Russian Co-Orbital" section of the SWF counterspace report: "Russia," Ch. 2 in *Global Counterspace Capabilities*, 2-2 to 2-14, and associated footnotes, Brian Weeden and Victoria Samson, eds., *Global Counterspace Capabilities: An Open Source Assessment*, (Washington, DC: Secure World Foundation (SWF), April 2021), https://swfound.org/media/207162/swf_global_counterspace_capabilities_2021.pdf

technologies, especially maneuver and target manipulation; the ability to routinely conduct such RPOs in higher orbits; the ability, highly likely in the near future, to field substantial numbers of hostile RPO-capable spacecraft relatively cheaply, due largely to miniaturization and economies of scale; and the fact of China's entry into the co-orbital ASAT club. The R-spacecraft considered in this report are not your *dedushka*'s space mines: the much finer manipulation of which new R-spacecraft are capable, as well as their ability to conduct serious hostile RPOs without generating debris, result in a substantially different threat profile than the earlier co-orbital ASATs.

(ii) *China's Program*

Numerous senior U.S. officials and intelligence agencies have issued public statements since at least 2018 expressing concern about China's growing ability to conduct targeted hostile RPOs.[344] For example, Gen. James Dickinson, the commander of U.S. Space Command, testified before the Senate in April 2021 that there was a serious and growing risk of China developing a "future system for grappling other satellites."[345] Such concerns are well-founded. China has been conducting RPO tests in orbit since at least 2008,[346]

---

344. Brian Chow, "Space Traffic Management in the New Space Age," *Strategic Studies Quarterly* vol. 14, no. 4 (Winter 2020): 97, https://www.airuniversity.af.edu/Portals/10/SSQ/documents/Volume-14_Issue-4/Chow.pdf#page=24 .

345. Gen. James H. Dickinson, "SUBJECT: Fiscal Year 2022 Priorities and Posture of the United States Space Command," United States Space Command, Presentation to the Senate Armed Services Committee, April 21, 2021, https://www.armed-services.senate.gov/imo/media/doc/Dickinson04.20.2021.pdf#page=6.

346. Brian Weeden, "Chinese Military and Intelligence Rendezvous and Proximity Operations Fact Sheet," Secure World Foundation (May 2022): 3, https://swfound.org/media/207367/swf-chinese-militarintel-rpo-may-2022.pdf#page=3.

and in an around Geosynchronous Orbit (GEO) since at least November 2016.[347] Over the past year, these fears have started to come to fruition.

Despite China's rapidly advancing space program, and the fact that a U.S. company—Space Logistics, a subsidiary of Northrop Grumman—carried out the first successful docking mission with another GEO satellite as early as February 2020,[348] some skeptics nonetheless continued to express doubt through late 2021 that China would have the ability or desire to attain the technical capability for such sophisticated co-orbital operations in the near future. Subsequent developments have rendered such skepticism untenable. In September 2021, the Shanghai Academy of Spaceflight Technology (SAST) officially unveiled the "Supplemental Service Aircraft," a spacecraft for commercial servicing that "takes over attitude orbit control of GEO satellites at the end of their lifetime to extend their in-orbit service life."[349] This reportedly comes on the heels of successful on-orbit tests of docking and refueling with the Tiangong-2 space laboratory.[350] Then, on October 23, 2021, China launched the Shijian-21 (SJ-21), a classified spacecraft also developed by the state-owned China Aerospace Science and

347. Weeden, "Chinese RPO Fact Sheet," 2.
348. Chris Gebhardt, "Northrop Grumman makes history, Mission Extension Vehicle docks to target satellite," NASA Spaceflight, February 26, 2020, https://www.nasaspaceflight.com/2020/02/northrop-grumman-history-mission-extension-vehicle-docks-satellite/.
349. CNSA Watcher [@CNSAWatcher], *[SAST's new supplemental service aircraft in #Zhuhai Airshow] It's a commercial service aircraft that takes over the attitude and orbit control of GEO satellite at the end of its life and prolongs its service life in orbit, helped by AI to identify target features and plan routes* [Tweet], Twitter, Sept. 10, 2021, https://twitter.com/CNSAWatcher/status/1436257779195555857 .
350. Global Times, "'Space oil tank truck' makes debut at Airshow China, provides energy for satellites' long-term operation," Global Times, Sept 29, 2021, https://www.globaltimes.cn/page/202109/1235426.shtml.

Technology Corporation (CASC)[351] which, according to Chinese state media Xinhua, is intended to "test and verify space debris mitigation technologies"—that is, the ability to close with, manipulate, and capture space objects—in GEO.[352] Beginning November 1, 2021 another object was catalogued maneuvering with it, akin to past Chinese ASAT tests.[353] Finally, in January 2022, the SJ-21 docked with a defunct BeiDou satellite and towing it above GEO.[354] While this mission constitutes a valuable reduction in orbital debris risk, it also demonstrates conclusively that China is capable of such operations, which alone is enough to warrant preparations so that should the worst-case scenario – to the authors, far more likely to be the reference-case scenario -- come to pass the U.S. is able to effectively deter, respond, and de-escalate the situation. Moreover, the facts that China released no details regarding this satellite or its mission and that there is evidence strongly suggestive of military involvement in its development and operation provide additional red flags.[355]

These developments culminated in Gen. Dickinson's testimony before Congress in March 2022 that "Shijian-17 and Shijian-21, which are satellites with robotic arm technology, could be used in a future system for grappling and disabling other satellites."[356]

---

351. Andrew Jones, "China's Shijian-21 towed dead satellite to a high graveyard orbit," SpaceNews, January 27, 2022, https://spacenews.com/chinas-shijian-21-spacecraft-docked-with-and-towed-a-dead-satellite/.
352. Andrew Jones, "China launches classified space debris mitigation technology satellite," SpaceNews, October 24, 2021, https://spacenews.com/china-launches-classified-space-debris-mitigation-technology-satellite/.
353. Andrew Jones, "An object is now orbiting alongside China's Shijian-21 debris mitigation satellite," SpaceNews, November 5, 2021, https://spacenews.com/an-object-is-now-orbiting-alongside-chinas-shijian-21-debris-mitigation-satellite/.
354. Harrison, "Space Threat Assessment 2022," 24.
355. Jones, "China's Shijian-21 towed."
356. Debra Werner, "An In-Orbit Game of Cat and Mouse: Close approaches prompt calls for communications and norms," SpaceNews, June 16, 2022,

Finally, in early 2022, the U.S. attempted to conduct close-approaches to two new Chinese technology testbed satellites, the Shiyan-12 (01) and (02), presumably to examine them; both satellites then maneuvered away from the U.S. satellite successfully, underscoring that China not only has sufficient capability for sophisticated and coordinated RPOs in GEO, but that they also already have sufficient high-fidelity timely space situational awareness (SSA) capabilities to monitor and swiftly respond to even small changes in the environment as far out as GEO.[357]

The available evidence is strongly suggestive of a dedicated ASAT program. Nevertheless, some influential space experts have noted—quite correctly—that these tests have plausible peaceful purposes, and are in many respects quite similar to those conducted by the United States. It is the assessment of the authors that active development—and bold use—of offensive R-spacecraft is in direct alignment with China's declared goals, strategic doctrine, observed behavior, and the available evidence. However, this report will not attempt to prove conclusively that these activities are an attempt to develop a dedicated rendezvous ASAT capability, because doing so is unnecessary to this paper's core purpose and premise: to identify and evaluate the *possible* threat and to recommend appropriate solutions. It is impossible to know intent with certainty. Yet even if it *were* somehow possible to be entirely certain that these capabilities were currently intended *solely* for peaceful applications, China—or any other country—could at any time in the future decide to launch a program adapting these technologies, techniques, and lessons-learned for counterspace use. Indeed, they are far more likely to calculate that doing so is in their best interest if the U.S. fails to enact and clearly communicate adequate countermeasures! Given the years of time lag involved with readying effective legal,

---

https://spacenews.com/an-in-orbit-game-of-cat-and-mouse-close-approaches-prompt-calls-for-communications-and-norms/.
357. Jones, "China's Shijian-21 towed."

policy, and technical counters to offensive RPOs, this alone is reason enough to prepare now, ensuring adequate defense against contemporaneous R-spacecraft threat.

Additionally, the severity of the potential consequences involved demands readiness. If the U.S. fails to prepare adequately for this particular threat vector, the results could be catastrophic, ranging from serious economic harm, to undercutting the liberal international order, and even to escalation to a large-scale conventional or nuclear conflict.

(iii) Potential Consequences

A defining feature that sets R-spacecraft apart from other ASAT threat vectors is that they may be "pre-positioned" in close proximity to their targets. This ability to loiter nearby has a number of serious implications, explored in detail below. At present, there are no explicit constraints placed on such close-approaches either by prevailing interpretations of international law or by the unilateral declaratory policy of any country, the United States included. Therefore, a potential adversary may, if they so choose, station *any number* of satellites *however close* they wish to *any number* of potential "targets" simultaneously. This is true regardless of the value or vulnerability of the targets in question—it matters not whether they handle television broadcasts or nuclear launches. The upshot is that R-spacecraft provide a unique means of interfering with or destroying—or of *threatening* to interfere with or destroy—a large number of satellites, and of doing so *suddenly*, *simultaneously*, and without concern for how *high* the target's orbit.

Unless the U.S. moves soon to adopt properly tailored policy, legal, and technical countermeasures, this report predicts that the advent and proliferation of these new and more sophisticated RPO capabilities will pose three primary threats to U.S. interests and the stability of the international system by the latter half of this decade. In descending order of severity, these three threats are: 1.) a threat to highly sensitive strategic-level assets in GEO, and especially space-based components of the U.S.'s nuclear weapons architecture; 2.) a threat to operationally significant GEO military and intelligence satellites and even commercial satellites; and 3.) a variety of risks to assets in LEO. All three of these threats warrant serious consideration and preparations to counter; nevertheless, the former two are categorically the most urgent. There is some room for debate regarding the probability that they will manifest: there are sound reasons to believe that the likelihood may be high, but the confidence interval of any specific prediction regarding other countries' behavior five years from now is inherently low. However, the sheer magnitude of the potential impact demands that the United States prepare in advance. Doing so is the only way to ensure that either an adversary never attempts such an operation in the first place, or that if they do, that the U.S. has response options at its disposal which are simultaneously both effective and de-escalatory or stability-enhancing.

Threat to Strategically Vital Assets

While satellites in Low Earth Orbit (LEO) have become increasingly vulnerable over the past three decades,[358] most of the U.S.'s highest-value assets have thus far remained largely free from threat on account of their distance, their hardening, and other factors. This is particularly true of the U.S.'s nuclear and conventional communications satellites at GEO, nearly 36,000km

---

358. See generally, e.g., Robin Dickey, "The Rise and Fall of Space Sanctuary in U.S. Policy," Center for Space Policy and Strategy, (Arlington, VA: The Aerospace Corporation, September 2020), https://aerospace.org/paper/rise-and-fall-space-sanctuary-us-policy .

above the Earth's surface, and early warning satellites that are stationed in GEO and Highly Elliptical Orbit (HEO).[359]

While a small handful of potential ASAT threats to GEO targets exist, all of them are heavily limited in various ways, and will remain so for the foreseeable future. China's successful test-launch of a direct-ascent DA-ASAT to an apogee of ~30,000km[360] is likely the best-known such capability. This Chinese missile[361] has probably not yet reached operational status, though it likely will between now and 2025.[362] However, even if China were to acquire

---

359. The subset of such satellites under discussion here are housed in GEO. See *supra* at David, "On-Orbit Servicing." GEO is a unique circular orbit with fixed altitude 35,786km in the equatorial plane. David Wright, Laura Grego, and Lisbeth Gronlund, *The Physics of Space Security: A Reference Manual* (Cambridge, MA: American Academy of Arts and Sciences, 2019), 43, https://aerospace.csis.org/wp-content/uploads/2019/06/physics-space-security.pdf#page=55 . Regarding their lack of vulnerability even to cyber threat vectors, see: David A. Deptula, William A. LaPlante, and Robert Haddick, "Modernizing U.S. Nuclear Command, Control, and Communications," (Arlington, VA: The Mitchell Institute for Aerospace Studies and MITRE Corporation, February 2019), http://docs.wixstatic.com/ugd/a2dd91_ed45cfd71de2457eba3bcce4d0657196.pdf .

360. China claims that it was a sounding rocket, but the evidence to the contrary is overwhelming. "China," Section 1 in Weeden and Samson, "*Global Counterspace Capabilities 2022*," 03-12 and associated footnotes. See also David Wright, "How High Did China's May 2013 Launch Go?" (Cambridge, MA: Union of Concerned Scientists (UCS), March 2014), https://allthingsnuclear.org/dwright/how-high-did-chinas-may-2013-launch-go/ .

361. Assessed by U.S. officials to be the DN-2. "2015 Annual Report to Congress," (Washington, DC: U.S.-China Economic and Security Review Commission (USCC), November 2015), 292-294, https://www.uscc.gov/sites/default/files/annual_reports/2015%20Annual%20Report%20to%20Congress.PDF ; Bill Gertz, "China Conducts Test of New Anti-Satellite Missile," (Washington, DC: The Washington Free Beacon, May 2013), https://freebeacon.com/national-security/china-conducts-test-of-new-anti-satellite-missile/. This is probably correct, as it appears to have been a test of a rocket component of a new direct-ascent ASAT weapons system based on an existing road-mobile ballistic missile, and had a flight trajectory far in excess of the pre-existing SC-19/DN-1's probable capability. See e.g., Weeden and Samson, "*Global Counterspace Capabilities 2021,*" 1-11 to 1-18, and footnotes.

362. Ibid.

a fully operational DA-ASAT capability to GEO, the threat posed to U.S. nuclear command, control and communications (NC3) satellites will pale in comparison to that from pre-positioning of R-spacecraft by mid to late 2020s.

There are several primary reasons for this. First, consider the speed and suddenness of attack. As Gen. Hyten has noted publicly, a DA-ASAT would take approximately 4-6 hours to reach a target in GEO, much longer than the approximately 6 minute lower-bound to strike a LEO target.[363] This buys precious time to avoid the impact.[364] Even more importantly, it guarantees that any such attack will be clearly visible well before it arrives.[365] To this transit time to target, add the lengthy and highly visible preparations required prior to launch.[366] This is particularly salient and unavoidable in this context, because reaching any GEO target will require use of a launch pad for the foreseeable future, of which China has only a handful.[367] At any point during this process, the missile and/or

---

363. Phillip Swarts, "Space Wars: The Air Force Awakens," (Air Force Times, February 2016), https://www.airforcetimes.com/news/your-air-force/2016/02/15/space-wars-the-air-force-awakens/. See also Wright, Grego, and Gronlund, *Physics of Space Security*, 156:

> [I]t could be hours before a ground-based ASAT was in the proper position to attack a satellite in low earth orbit," but even if this were not the case because the DA-ASAT was already positioned such that it can launch directly from where it is stationed into an intercept trajectory with the desired target—the absolute best-case scenario for a prospective attacker—it would then "reach a satellite in low earth orbit in a matter of minutes [but still] take several hours to reach a satellite in geosynchronous orbit.

364. Ashton B. Carter, "Satellites and Anti-Satellites: The Limits of the Possible," *International Security* 10, no. 4 (Spring 1986): 80-84, http://www.jstor.org/stable/2538950. For an excellent primer on the physics involved in on-orbit maneuver (and all other facets of the counterspace issue), see generally Wright, Grego, and Gronlund, *Physics of Space Security*.
365. Ibid.
366. See Wright, Grego, and Gronlund, *Physics of Space Security,* 152.
367. Ibid; Geoffrey Forden, "How China Loses the Coming Space War (Pt. 2)," (Wired, January 2008), https://www.wired.com/2008/01/inside-the-ch-1/; Chinese TEL-capable missiles have nowhere near the required range – "Missiles

launcher could be destroyed.[368] Even if it is not, these steps take time, which can enable avoidance maneuvers, the readying and deployment of terrestrial fill-ins, the launch of replacement satellites, the swift imposition of proportionate costs, and so on.[369] Additionally, because these factors functionally throttle the number of DA-ASATs able to be launched over a given period of time, they directly limit how many GEO satellites can be destroyed over that period, as well as determine how long the time lag is between each intercept or group of intercepts.

In contrast, the unrestricted placement of R-spacecraft near numerous targets would enable an attacker to, at will, interfere with or destroy *many satellites simultaneously*; to be confident of their ability to do so; and to implicitly and credibly *threaten* to do so. Moreover, they can even benefit from the implicit coercive pressure of a looming threat while claiming to be operating peacefully. Comparable effects are highly unlikely to be achieved by almost[370]

of China," Missile Threat (Washington, DC: CSIS), https://missilethreat.csis.org/country/china/ (Note: Typically, a ballistic missile apogee is roughly half its horizontal range - see: http://swfound.org/media/167224/Through_a_Glass_Darkly_March2014.pdf ; Mostly from Xichang, see "China," Ch.1 of Weeden and Samson, "*Global Counterspace Capabilities 2021.*"

368. Jaganath Sankaran, "Limits of the Chinese anti-satellite Threat," *Strategic Studies Quarterly* (Winter 2014): 25, https://www.airuniversity.af.edu/Portals/10/SSQ/documents/Volume-08_Issue-4/Sankaran.pdf .

369. Sankaran, "Limits of the Chinese ASAT Threat," 24-25.

370. Perhaps the most prominent threat vector not meaningfully assessed in this report due to length constraints is cyber, whether targeted at the supply chain, satellite bus, ground control station, operator, or any other component. The authors have done work in that area elsewhere, and while there is much of value to be said there, none of it undercuts the importance of this paper's findings. This is so for a range of reasons, but in lieu of an exhaustive footnoted list, the following quote from Gen. Hyten will suffice for purposes of this paper: "[NC3] is an old system. But it's very secure. If you want a cyber-secure system, build it in the 1960s." John Hyten, interviewed by Sandra Erwin, "Q&A: Air Force Gen. John Hyten says U.S. space strategy, budget moving 'down the right path',"

any other means for the foreseeable future, and as a result R-spacecraft pose a unique threat in the narrow but critical context of non-dispersible, very-high-value assets in Medium Earth Orbit (MEO), GEO, and HEO. Offensive cyber capabilities are an important possible exception, but are deliberately bracketed from this report as they deserve thorough assessment and countering in their own right, are certainly no reason to not adequately deal with *this* critical threat vector, and in any case are substantially less threatening in the specific context of nuclear command, control, and communications systems because of their old age and analog design.[371]

The ability of R-spacecraft to strike suddenly and without warning produces a threat of the proverbial Swords of Damocles. In this way, the pre-positioning of R-spacecraft is likely to pose a unique and serious risk to strategic stability, especially during crises.[372] In 1985,

---

(SpaceNews, April 2018), https://spacenews.com/qa-air-force-gen-john-hyten-says-u-s-space-strategy-budget-moving-down-the-right-path/ .

371. Cyber is perhaps the most prominent threat vector not meaningfully assessed in this report due to length constraints, whether targeted at the supply chain, satellite bus, ground control station, operator, or any other component. The authors have done work in that area elsewhere, and while there is much of value to be said there, none of it undercuts the importance of this paper's findings. This is so for a range of reasons, but in lieu of an exhaustive footnoted list, the following quote from Gen. Hyten will suffice for purposes of this paper: "[NC3] is an old system. But it's very secure. If you want a cyber-secure system, build it in the 1960s." John Hyten, interviewed by Sandra Erwin, "Q&A: Air Force Gen. John Hyten says U.S. space strategy, budget moving 'down the right path'," (SpaceNews, April 2018), https://spacenews.com/qa-air-force-gen-john-hyten-says-u-s-space-strategy-budget-moving-down-the-right-path/.

372. As Ashton Carter (25th Secretary of Defense, serving Director of the Belfer Center for Science & International Affairs at Harvard Kennedy School as of this writing) noted, the task is not to identify the "technologies of ASAT attack" or their capabilities in a vacuum, but rather "identifying the plausible scenarios for their use." This is crucial: one "must not be alarmed by the mere capability to attack a satellite if a plausible reason for doing so cannot be identified." For this reason, "a sensible winnowing of real security problems from phony 'windows of vulnerability' is a major research problem." This paper contends that pre-positioning of prospective co-orbital ASATs in this narrow context constitutes an

Albert Wohlstetter and Brian Chow introduced the self-defense zones, allowing a zone's defender to "have the right to inspect, expel or otherwise render harmless any invaders".[373] Ashton Carter argued in 1986 that "without agreed or tacitly accepted keep-out zones [same concept as self-defense zones], space mines could easily develop into the single most pernicious threat to a stable military regime in space."[374] At this time and in the context of critical GEO satellites, at least, he appears quite correct.

Given the risks involved, one might expect that China would be self-deterred from or uninterested in making such threats. This is certainly possible, but the authors judge it unlikely. A comprehensive assessment of China's likely behavior is outside the scope of this report, and covered at length elsewhere. However, for purposes of this report, it is sufficient to establish that the scenario of attacking critical GEO satellites is *plausible*—which, unfortunately, is an understatement.

The risk arises in part from the fact that China and the U.S. notably lack the shared context and mutual understanding around nuclear and crisis stability issues that characterized the relationship between the U.S. and the Soviet Union during the Cold War. China appears to systematically discount the possibility of nuclear escalation, generally believing that even a single nuclear weapon is sufficient to establish a credible deterrent, and by extension is risk-acceptant in the extreme in ways that line up quite dangerously with the U.S.'s

---

example of the former rather than the latter. Carter, "Satellites and Anti-Satellites," 80.

373. Albert Wohlstetter and Brian G. Chow, "Arms Control That Could Work (1985)," in *Nuclear Heuristics: Selected Writings of Albert and Roberta Wohlstetter*, edited by Robert Zarate and Henry Sokolski, January 2009, 501-505, https://www.files.ethz.ch/isn/95998/2009_01_NuclearHeuristics.pdf#page=513 .

374. Carter, "Satellites and Anti-Satellites," 87.

quite different perspective on nuclear strategy.[375] At the same time, Chinese strategic thought and military doctrine heavily prioritizes strategic surprise, keeping one's intentions opaque, maintaining the initiative, and striking first.[376] As a result, in sharp contrast to how

375. The U.S. and Russia have a shared (albeit imperfect) understanding of nuclear issues, developed over the course of decades and with some very significant near-misses, which centers strategic stability and mutually-assured destruction (MAD). China, in contrast, sees risk-manipulation, opacity, strategic surprise, seizure of the initiative, and domination of the information environment as essential to success. This belief is consistently evident in PLA doctrine, white papers, statements by senior PLA officials, decisions regarding what capabilities to acquire, and a range of other sources. See: Laura Grego, "Space and Crisis Stability," (Union of Concerned Scientists, 2018), 31, https://archive.law.upenn.edu/live/files/7804-grego-space-and-crisis-stabilitypdf.
It is therefore somewhat unsurprising that many senior PLA officials appear to believe that interfering with U.S. NC3 nodes or conducting "limited" strikes on early warning satellites might be an effective means of convincing the U.S. to back down or deterring further escalation. See: ibid.
Similarly, we may expect China to probably conclude that keeping their intentions opaque while positioning to hold critical assets at risk is an effective means of maintaining the strategic initiative and gaining leverage over the U.S. More generally, China is likely to be systematically overconfident in their ability to manage escalation risk, and, most concerningly, to believe that inflicting greater costs will deter further escalation. This is a concern in the cyber domain and terrestrially, and is no less worrisome in this high-stakes counterspace context. For more, see: Caitlin Talmadge, "Would China Go Nuclear? Assessing the Risk of Chinese Nuclear Escalation in a Conventional War with the United States," *International Security* 41, no. 4 (Spring 2017), 50–92, https://direct.mit.edu/isec/article-abstract/41/4/50/12156/Would-China-Go-Nuclear-Assessing-the-Risk-of?redirectedFrom=fulltext. See also: Science of Military Strategy; and https://www.uscc.gov/sites/default/files/2021-02/Fiona_Cunningham_Testimony.pdf.
376. Regarding Chinese motivation and willingness to undertake such strikes, it is important to note that:
Chinese military writings overall place a heavy emphasis on gaining the initiative at the outset of a conflict, including during the deployment stage. Looking at the 1991 Gulf War[…] Chinese military analysts assess that the PLA cannot allow the U.S. military to become fully prepared lest they cede victory… [and therefore] China should 'strive to attack first at the campaign and tactical levels in order to maintain the space battlefield initiative.' They also argue that fighting a quick war is one of the 'special characteristics of space operations' and that a military should 'conceal the concentration of its forces and make a decisive large-scale first strike.' "China," Ch. 1 in "*Global Counterspace Capabilities 2021*," 30. [Internally quoting: Jiang Lianju and Wang Liwen

the United States might act if similarly-positioned, China might well be inclined to attack one or more very-high-value assets in the early stages of a conflict—or even before any conflict has actually begun—under the premise that bloodying the U.S.'s nose in this manner would convince it of China's seriousness and thereby induce it to back down. Senior U.S. officials have reported troubling informal dialogues with their Chinese counterparts in which the latter have specifically flagged early strikes on U.S. early warning satellites as one potential mechanism for signaling their seriousness, and dismissively brushed off the Americans' alarmed reactions as merely attempts to bluff or posture. Unfortunately, far from deterring the U.S., such a move might well invite a massive response. Depending on the targets, nuclear escalation is theoretically possible, and technically consistent with official U.S. doctrine.[377] Far more likely, however, the U.S. would respond with conventional strikes against key Chinese targets—possibly even on the Chinese mainland, with nuclear command-and-control nodes and missile sites being particularly likely targets. Such a scenario might quickly escalate to a broader conflict—and may even ultimately result in China crossing the nuclear threshold themselves.[378]

A more specific context in which R-spacecraft might conceivably be employed with disastrous consequences is a potential contingency

---

(Eds.), *Textbook for the Study of Space Operations* (空间作战学教程), (Beijing, China: Military Science Publishing House, 2013), 52, 142-143.]

377.Ankit Panda, "Space-Based Nuclear Command and Control and the 'Non-Nuclear Strategic Attack'," (The Diplomat, April 2020), https://thediplomat.com/2020/04/space-based-nuclear-command-and-control-and-the-non-nuclear-strategic-attack/ .

378. Escalation risk is systematically underestimated in the China context. Crossing the nuclear threshold is a real possibility even if (as is probable) neither side enters the crisis intending that result. See: Talmadge, "Would China Go Nuclear?"

over Taiwan.[379] This scenario is chosen in part because it is alarmingly plausible, arguably the likeliest site for Sino-American war within the next three to seven years; but primarily, and much more importantly, because of what it reveals about the unique threat characteristics of pre-positioned R-spacecraft.

Chinese attempts to seize Taiwan are becoming more thinkable by the day. On July 1, 2021, the 100th birthday of the Chinese Communist Party (CCP) President Xi Jinping publicly vowed that China will "advance peaceful national reunification"[380] with Taiwan. On October 9, 2021, Xi declared that "the historical task of the complete reunification of the motherland must be fulfilled, and will definitely be fulfilled."[381] Meanwhile, Gen. Mark Milley, Chairman of the Joint Chiefs of Staff, testified before Congress on June 17, 2021 that the People's Liberation Army (PLA) has accelerated the pace of its military modernization programs—specifically to meet a goal declared by President Xi of attaining the capability by 2027 to successfully invade and seize Taiwan by force.[382] (This is an acceleration from the PLA's previously-declared timeline of 2035.) This warning echoes congressional testimonies from numerous other U.S. officials, including the

379. Please note that comprehensive evaluation of a Taiwan contingency is beyond the scope of this report, and any disagreements with either the likelihood or details of the scenario do not undercut the core point being made. Enough will be said to demonstrate the bare plausibility of the scenario, but the primary focus is on what the hypothetical reveals regarding the unique characteristics of the rendezvous spacecraft threat vector.

380. "Full text of Xi Jinping's speech on the CCP's 100th anniversary," Nikkei Asia, July 1, 2021, https://asia.nikkei.com/Politics/Full-text-of-Xi-Jinping-s-speech-on-the-CCP-s-100th-anniversary .

381. Vincent Ni and agencies, "Xi Jinping vows to fulfill Taiwan 'reunification' with China by peaceful means," The Observer, Taiwan, October 9, 2021, https://www.theguardian.com/world/2021/oct/09/xi-jinping-vows-taiwans-reunification-with-china-will-be-fulfilled .

382. Sam LaGrone, "Milley: China Wants Capability to Take Taiwan by 2027, Sees No Near-term Intent to Invade," USNI News, June 23, 2021, https://news.usni.org/2021/06/23/milley-china-wants-capability-to-take-taiwan-by-2027-sees-no-near-term-intent-to-invade.

former head of U.S. Indo-Pacific Command, Adm. Phil Davidson,[383] and current commander, Adm. John Aquilino.[384]

At the same time, the PLA has centralized their space (along with cyber and missile) capabilities within the Strategic Support Force (SSF). China scholars have noted that in any attempted invasion of Taiwan the PLASSF would almost certainly be specifically tasked with leveraging these capabilities to "*undermine the U.S. military's capability [and will] to deploy force…within a relevant timeframe*" in order to "*prevent or preempt the potential for American intervention.*"[385]

R-spacecraft, especially if specifically designed for co-orbital ASAT use, might well prove attractive as a means of achieving this goal. By pre-positioning such satellites in close proximity to numerous strategically vital U.S. satellites during peacetime, China would gain a unique ability to conduct ASAT attacks at will, with no prior warning, against as many targets as desired—as well as to threaten to do so, either explicitly or implicitly. (Depending on the weapon system, this could be permanent or reversible, and debris-producing or not.) This would uniquely endanger a number of key U.S. constellations which are otherwise likely to be relatively safe, including those responsible for missile early warning, nuclear and conventional communications, and, as

383. Mallory Shelbourne, "Davidson: China Could Try to Take Control of Taiwan In 'Next Six Years'," USNI News, March 9, 2021, https://news.usni.org/2021/03/09/davidson-china-could-try-to-take-control-of-taiwan-in-next-six-years.
384. Mallory Shelbourne, "Military Takeover of Taiwan Is Top Concern for INDOPACOM Nominee Aquilino," USNI News, March 23, 2021, https://news.usni.org/2021/03/23/military-takeover-of-taiwan-is-top-concern-for-indopacom-nominee-aquilino.
385. Elsa B. Kania and John Costello, "Seizing the commanding heights: the PLA Strategic Support Force in Chinese military power," *Journal of Strategic Studies*, vol. 44, no. 2 (2021): 218-264, DOI: 10.1080/01402390.2020.1747444.

discussed *infra* at next subsection on the Global Positioning System (GPS).

This would generate serious escalation risks. China might well assess that threatening such assets, or even striking them in the early stages of a Taiwan Strait contingency if U.S. intervention appears likely, would give the U.S. pause about intervening on Taiwan's behalf. Should they prove to be correct, U.S. failure to respond would likely do serious damage to U.S. credibility and the global alliance system it underpins, with implications for U.S. influence, the non-proliferation regime, deterrence in key regions, and more.[386] The alternative, however, may be even worse: if China makes this gamble and is proven wrong, the resultant escalation could be swift and unpredictable, resulting in, at best, a conventional great power conflict with magnified costs in blood and treasure due to the need for the U.S. to fight without certain key satellites; and, at worst, scenarios for nuclear exchange.

Threat to Strategically and Operationally Significant Assets

For the same reasons and in the same ways that mass pre-positioning of R-spacecraft poses a serious threat to space-based elements of the U.S.'s nuclear weapons architecture, they also endanger a range of space assets important for conventional military and intelligence operations. A comprehensive net-assessment of these capabilities and the military-operational implications of their loss falls well

386. Hal Brands, Eric Edelman, and Thomas G. Mahnken, "Credibility Matters: Strengthening American Deterrence in an Age of Geopolitical Turmoil," (Washington, DC: Center for Strategic & Budgetary Assessments (CSBA), 2018), https://csbaonline.org/uploads/documents/Credibility_Paper_FINAL_format.pdf .

beyond the scope of this report. However, this report will discuss a single potential target set of particularly high salience, with an eye toward illuminating the uniqueness of the threat posed by R-spacecraft: the U.S.'s Global Positioning System constellation, better known as GPS, which resides in MEO.

Loss of GPS would substantially degrade the U.S.'s warfighting capability in Northeast Asia, especially if terrestrial positioning, navigation and timing (PNT) backups prove less capable than DOD hopes.[387] At absolute minimum, it certainly would unleash a wave of secondary repercussions upon the U.S. economy and various civilian applications.[388] There is also the question of response: it is difficult to imagine that the reaction to sudden loss of the GPS constellation would be a measured one.

Unfortunately, the same two factors that make it so costly—the potential impact on the U.S.'s over-the-horizon warfighting capability, and the U.S.'s strong attachment to and societal dependence on GPS—also make it an exceptionally attractive target for China. And, none of the possible limiting factors that could save NC3 satellites from being targeted apply here: GPS serves as a direct enabler of conventional military operations, and the potential for space debris is much less likely to be a constraint in MEO than in GEO as the odds (and potential costs) of collateral damage are many orders of magnitude smaller in the former orbit than the latter.

---

387. James Hasik, "The American Military's Greatest Vulnerability: No GPS in a War," The National Interest, October 18, 2016, https://nationalinterest.org/blog/the-buzz/the-american-militarys-greatest-weakness-no-gps-war-18085 .

388. Reporter quoting numerous senior US military & intelligence personnel, Garrett M. Graff, "The New Arms Race Threatening to Explode in Space," (Wired, June 2018), https://www.wired.com/story/new-arms-race-threatening-to-explode-in-space/; Regarding the importance of GPS, see "Thirteenth Meeting Agenda of the National Space-Based Positioning, Navigation, and Timing Advisory Board," (2014), https://www.gps.gov/governance/advisory/meetings/2014-06/minutes.pdf .

Fortunately, the GPS constellation is extremely well-protected against counterspace threats—*except* for that posed by pre-positioned R-satellites. Despite a constant stream of commentary occasionally verging on the hysterical, it is exceptionally challenging to meaningfully degrade GPS to any strategically- or even operationally-impactful extent, because of its nature as a distributed and redundant system, and in light of supplemental hardening efforts, preparedness to swiftly reconstitute missing pieces of the constellation, and so on.[389] The U.S. currently has some 31 operational GPS satellites in orbit, with 24-27 of them active at any one time.[390] Each satellite orbits twice per day,[391] and approximately 9 are over China at any one time.[392] Therefore, one

---

389. See generally, Brandon Davenport and Rich Ganske, "'Recalculating Route': A Realistic Risk Assessment for GPS," (War on the Rocks, March 2019), https://warontherocks.com/2019/03/recalculating-route-a-realistic-risk-assessment-for-gps/.

Wright, Grego, and Gronlund, *Physics of Space Security*, 165-166:

> Degradation or loss of the signal without prior planning could seriously compromise military and economic life, but "[a]n important feature of the GPS constellation is that its ability to provide navigation information would degrade gradually, rather than catastrophically, under an attack." Moreover, "[t]he GPS system…has some innate protections from attack. The satellites operate at a high altitude, which puts them out of reach of ground-based kinetic energy attacks using modified short-range or intercontinental ballistic missiles….Blinding and dazzling would not affect the ability of the satellites to provide navigation information since doing so requires no optical sensors. The satellites themselves are designed to withstand heat imbalances from distant laser attacks.[Citation removed] Moreover, the satellites are in six separate orbital planes and their spacing within those orbits is such that a single ASAT would have difficulty targeting more than one satellite.

[Internal citation for the claim that GPS is hardened against DEW-induced heat imbalances: Carter, "Satellites and Anti-Satellites," 90.]

390. "GPS: Space Segment," GPS.Gov, https://www.gps.gov/systems/gps/space/#:~:text=The%20GPS%20space%20segment%20consists,%2C%2095%25%20of%20the%20time .

391. Orbital period of ~12 hours sidereal time. See "GPS Space Segment," (ESA, 2011), https://gssc.esa.int/navipedia/index.php/GPS_Space_Segment .

392. Forden, "How China Loses (Pt. 2)."

would need to destroy or disable about 10 satellites to achieve a sustained GPS outage,[393] but even that would only achieve outages for a period of about 8 hours per day.[394]

In the Appendix and a more recent paper,[395] the authors assessed that China will likely have on the order of 200 R-spacecraft for ASAT purposes around 2026. Then, for example, China will be able to place R-spacecraft in orbits closely matching every single satellite in the GPS constellation. Such pre-positioning *would* enable China to destroy or disable an operationally- and strategically- impactful percentage including close to 100% of the constellation at-will: suddenly, simultaneously, and without adequate warning. Moreover, a fleet of 200 R-spacecraft would be enough to disable most, if not all, of the three-dozen or so GPS satellites with plenty left for other critical satellites,[396] whether military or commercial and whether

---

393. Ibid.

394. Wright, Grego, and Gronlund, *Physics of Space Security*, 165-166:

> Between five and eight satellites are visible from any point on the Earth at all times. However, only four satellites are required to be in view to provide [PNT]…. If the user can receive signals from more than four satellites, the accuracy improves…. [GPS] would degrade gradually, rather than catastrophically, under an attack. For example, the minimum requirement of four satellites would be available for almost the entire day to a user in Beijing even if six satellites were lost, even if those six were chosen to give the greatest loss of service for that location. [Citation removed]

[Internal citation for the modeling of GPS availability in the China context, assuming destruction of six satellites: Geoffrey Forden, "Appendix D: Sensitivity of GPS Coverage to Loss of One or More Satellites," Ensuring America's Space Security (Washington, D.C.: Federation of American Scientists, 2004), https://pubs.fas.org/_docs/10072004164336.pdf] See also: Sankaran, "Limits of the Chinese ASAT Threat"; and Forden, "How China Loses (Pt. 2)."

395. Brian G. Chow and Brandon W. Kelley, "Op-ed | Russian Invasion of Ukraine Reinforces the Urgency of Fixing U.S. Satellite Vulnerability by 2027," SpaceNews, March 8, 2022, https://spacenews.com/op-ed-russian-invasion-of-ukraine-reinforces-the-urgency-of-fixing-u-s-satellite-vulnerability-by-2027/.

396. Brian Chow and Brandon Kelley, China's Anti-Satellite Weapons Could Conquer Taiwan – Or Start a War, The National Interest, August 21, 2021, https://nationalinterest.org/feature/china%E2%80%99s-anti-satellite-weapons-

U.S. or its partners'. Thus, ultimately, preventing such satellites from being freely pre-positioned *en masse* in close proximity to our critical satellites during peacetime is necessary and sufficient to ensure the continued safety of GPS, as well as a number of other high-orbit military and intelligence assets which could be easily targeted in the same manner.[397]

As discussed in the Appendix, the authors assessed China's orders of battle in R-spacecraft for ASATs in 2027 and 2029 scenarios for NPEC's China Space Wargame in June 2021. These R-spacecraft were among other space threats used by both sides for counterspace purposes or deterrence against their use. While over two-dozen wargame participants have diverse experience and ideologies, they ended up to produce four takeaways,[398] all pertaining to the need of far better preparing against R-spacecraft ASATs, as they are considered it to be driving the outcome of the wargame.

Threats to Space Traffic and Tactically Significant Assets in Low-Earth Orbit

LEO includes all orbits below an altitude of 2000 km. The lower bound, without assistance, is ~120km for circular orbits and ~80km for elliptical orbits. Low LEO houses the majority of remote sensing

---

could-conquer-taiwan%E2%80%94or-start-war-192135 . This article showed the U.S. critical and vulnerable satellites that China might want to disable at the opening of a war. However, it should be expected that the 200 R-spacecraft may be ready as soon as 2026 instead of 2029 as in the article. See the Appendix for explanation regarding why these R-spacecraft will be available three years sooner.

397. For consideration of the quantity of R-spacecraft that China is likely to have available for use, please refer to the Appendix.

398. Henry Sokolski, editor, *China Waging War in Space: An After-Action Report*, Occasional Paper 2104, Nonproliferation Policy Education Center, August 2021, 4-5, https://npolicy.org/wp-content/uploads/2021/08/2104-China-Space-Wargame-Report.pdf .

satellites of various kinds—e.g., civil and commercial Earth observation (EO); atmospheric science; intelligence, surveillance, and reconnaissance (ISR); some navigation; and so on—because most such instruments require close proximity to Earth. It is also much more easily accessible than higher orbits, requiring substantially less energy to launch into. This renders satellite deployment much easier and less expensive, and makes reaching it with an ASAT capability (whether kinetic or otherwise, e.g., lasers or jamming) much easier for the same reason. Advances over the past fifteen years have reduced launch costs to LEO by orders of magnitude, and miniaturization of electronics has enabled the inexpensive production and lofting of constellations of large numbers of small satellites. The result has been an exponential boom in the number of satellites in orbit, as well as further increases in societal dependence on space for a wide variety of tasks. These trends are only set to accelerate for the foreseeable future.

The result is a LEO environment characterized simultaneously by increasing crowdedness, increasing importance, increasing levels of debris, and an increasing number of diverse actors. This produces a number of risks. As the orbital regime becomes more congested, both physical and radio frequency (RF) interference are likely to become increasingly commonplace, whether due to unavoidable accidents, carelessness, deliberate counterspace actions.

This intersects directly with R-spacecraft in several ways. First, more sophisticated RPO capabilities open a number of important doors for managing these challenges, such as the ability to physically change the orbits of satellites which are either non-responsive or lack functioning on-board propulsion, the ability to repair or refuel satellites *in situ* rather than launch a new one, and the ability to "clean up" space debris. This, in turn, however, creates an urgent need for frameworks to regulate such operations, and to manage the inevitable issues and tensions that will arise from them.

Absent such frameworks, it is easy to imagine a range of negative collateral consequences, whether due to direct and inadvertent collateral damage, misperception, or some other source.

Second, absent effective regulation, there is a real possibility of hostile RPOs being conducted in LEO. Given the much lower stakes and lessened vulnerability and impact typically involved in the LEO context as compared to the critical GEO and HEO satellites discussed above, the likely consequences differ dramatically. Two primary risks are likely to arise from offensive RPOs in low orbit. First, there is the possibility of damage to or interference with satellites of *tactical* significance; as discussed above, a great many assets call LEO home. This, however, is both not all that different from any other (i.e., non-rendezvous) ASAT threat to LEO satellites, and is limited considerably in likelihood and potential consequence by the ease of increasing proliferation of satellite constellations, and of reconstituting missing assets relatively quickly and cheaply where necessary. The second risk is much more alarming, and that is so-called "grey zone" use of LEO RPOs—the likelihood that R-satellites will be employed in ways that are aggressive and impactful, but fall well short of an armed attack, and are therefore difficult to respond to effectively and proportionately. In particular, operations targeting the assets of companies, civil agencies, or international allies and partners might well prove attractive to China (or others) as a means of undercutting international norms, provoking disagreements regarding the appropriate response, signaling their displeasure with a given action, or for some other reason.

A more comprehensive treatment of these issues by the authors is forthcoming elsewhere. However, this survey is sufficient for purposes of this report, because it clearly establishes two key points. First, that a meaningful RPO-related threat in LEO is emerging, and will become increasingly salient as the decade wears on. Thus, there is a pressing need to pursue internationally agreed-upon "rules of the road" for so-called "space traffic management" (STM), as well as to pursue unilateral best-practices both to help ensure the efficacy of this international regime and to mitigate the consequences should a potential adversary decide not to comply with it. Second, that while this concern is important, it is categorically distinct from the threat posed in higher orbits. The latter is not only much more severe and pressing, but also simply has very different characteristics than the former, and therefore should be evaluated separately and warrants an entirely different set of tailored solutions.

B. Lasers

(i) Threat Profile

The second threat selected by this report for evaluation over the rest of this decade is ground-based lasers (GBLs). Lasers, whether ground-based or airborne (ABLs), and other directed-energy weapons (DEWs) are growing increasingly capable and prolific. Generally speaking, lasers are of significant military value for a few reasons. First, they enable highly precise targeting and localized effects. While lasers of sufficient intensity are capable of destroying a target, laser beams of lower power may be employed to more granular effect, such as to cause reversible and non-destructive interference, e.g., by dazzling a sensor, or to permanently damage or destroy only a specific component, reducing a target's capacity or even taking it out of commission entirely all without generating debris in the process. Second, lasers cost much less per shot compared with long-range conventional munitions such as precision-guided missiles. Third, lasers can potentially enable engagement with many targets at long-range in rapid succession; contrast this with, for example, the logistical challenges inherent in launching large numbers of

DA-ASAT missiles on the right orbital trajectories over a short time window.

Both the U.S. and other countries have made major advances in laser technologies over the past decade. The primary constraint on laser capability at present is energy storage and generation: the physics of projecting light over long distances through the atmosphere in sufficient concentration to generate kinetic effects demand high energy throughput. While breakthroughs continue to be made at a rapid pace, there remains significantly further left to go. Nevertheless, assessment of China's current GBL program reveals a systematic effort to develop increasingly-capable counterspace laser capabilities, and there are clear signs that those efforts are bearing fruit. The United States should expect that China's laser capabilities will continue to grow more sophisticated over the course of this decade and beyond, and therefore should carefully evaluate the nature of this emerging threat in order to identify the future characteristics of this threat vector, the potential impacts if it is left unaddressed, and what combination of unilateral and diplomatic countermeasures are warranted in response.

(ii) China's Program

Testimony by numerous U.S. officials reveals a longstanding pattern of China employing GBLs to "paint" U.S. satellites, and in some cases to dazzle or damage the optical sensors onboard. These capabilities are growing, however. The Defense Intelligence Agency (DIA) warned in January 2019 that China would likely field a ground-based laser weapon the subsequent year capable of damaging space-based sensors in low earth orbit.[399] As of this writing in mid-2022, there are at least five Chinese fixed GBL sites clearly attested in the open-source literature.[400] Unfortunately, little-to-no recent evidence currently exists in the public domain regarding the power level of these sites. It is publicly known that as early as 2005[401] these GBLs had a likely power level of about 50-100 kilowatts.[402] Given advances in the underlying technology since that time, this appears commensurate with the publicly-stated assessment of the DIA and other U.S. intelligence agencies that these lasers are already capable of dazzling and damage *sensors*, as well as public estimates by those agencies that Chinese GBLs are likely to be capable of damaging at least sensitive *external structures* (e.g. solar panels) on both optical and non-optical satellites across a range of altitudes in LEO by no later than the mid-to-late 2020s.[403]

In addition to these fixed higher-powered sites, which warrant primary consideration, China also currently operates two

---

399. Doug Messier, "DIA: China Has Ground-based ASAT Missile, Likely Working on Laser Weapon," Parabolic Arc, February 18, 2019, http://www.parabolicarc.com/2019/02/18/dia-china-groundbased-asat-missile-working-laser-weapon/.
400. Col. Vinayak Bhat, "These futuristic Chinese space denial weapons can disable or destroy opposing satellites," The Print, March 23, 2019, https://theprint.in/defence/these-futuristic-chinese-space-denial-weapons-can-disable-or-destroy-opposing-satellites/210212/.
401. The lack of recent data calls for the government to supply more data and other information about threat and especially responses as the public has the need to know and the desire to better participate in space policy deliberation and decisions.
402. Bill Gertz, "Satellite Photos Show Chinese Anti-Satellite Laser Base," The Washington Free Beacon, April 1, 2019, https://freebeacon.com/national-security/satellite-photos-show-chinese-anti-satellite-laser-base/.
403. "Challenges to Security in Space," U.S. Defense Intelligence Agency (DIA), January 2019, 20, https://www.dia.mil/Portals/110/Images/News/Military_Powers_Publications/Space_Threat_V14_020119_sm.pdf#page=20.

mobile[404] satellite laser ranging (SLR) stations with very low power of 3 to 60 watts.[405] As of 2014, there were about 50 such stations total among all countries worldwide.[406] These ranging systems are used to accurately determine the orbits of satellites and space debris, and currently have low probability[407] to cause permanent damage to a satellite sensor in low-orbit due to their low wattage. However, this could easily change over time, as advances continue in laser amplification and miniaturization of power sources. There is also a cooperative SLR station of China and Argentina[408] in San Juan, suggesting that China could eventually, if it chose to make this a priority, develop the capability to temporarily dazzle or permanently damage our satellite sensors via lasers from stations outside of mainland China.

(iii) *Potential Consequences*

Lasers are notable both for what they *can* do and for what they *cannot* do, especially relative to other ASAT threats including that posed by R-satellites. Presently, the operational and tactical impact of ground-to-space lasers is somewhat narrowly limited.

---

404. "Satellite laser ranging techniques and their applications," Chinese Academy of Sciences: Shanghai Astronomical Observatory, September 5, 2014, http://english.shao.cas.cn/rh/rgs/200809/t20080905_9006.html
405. Zhibo Wu et al., "The progress of laser ranging technology at Shanghai Astronomical Observatory," *Geodesy and Geodynamics* vol. 10, iss. 6 (Nov. 2019): 492-498, https://www.sciencedirect.com/science/article/pii/S1674984719300552.
406. CAS, "Satellite laser ranging."
407. Brian G. Chow and Henry Sokolski, "Op-ed | U.S. satellites increasingly vulnerable to China's ground-based lasers," SpaceNews, July 10, 2020, https://spacenews.com/op-ed-u-s-satellites-increasingly-vulnerable-to-chinas-ground-based-lasers/.
408. Yan-Ben Han et al., "Successful operation of a cooperative SLR station of China and Argentina in San Juan," *Chinese Science Bulletin* vol. 53 (2008): 2417–2420, https://link.springer.com/article/10.1007/s11434-008-0338-x .

As referenced above, lasers are highly capable of dazzling or damaging sensitive optical sensors, and as such can it difficult to gather high-resolution imagery of targets in the vicinity of the GBL. As they become increasingly capable, lasers also pose a steadily growing threat to other sensors and sensitive electronics, making them a menace to a broader set of targets in LEO. However, as GBLs increasingly grow powerful enough to disable and destroy *external structures* such as solar panels on which satellites depend (and, eventually, the satellites themselves directly) in low orbits, this will meaningfully impact the operating environment, particularly at the tactical level. The physics and economics of satellites in LEO make it unlikely that most low-orbit satellites will be hardened against directed-energy weapon (DEW)-induced heat imbalances; the cost-exchange ratio heavily favors the attacker. As a result, China may attain the ability to take out a meaningful number of satellites in LEO, from almost any angle, swiftly, suddenly/with little to no notice, and crucially while producing little to no debris, making the capability much more usable. That being said, as discussed *infra* at Section III, proliferated constellations of satellites, and secondarily the ability to rapidly reconstitute critical assets, place this threat firmly in the category of "impactful, but non-catastrophic." Pairing these defenses with the other tailored unilateral and diplomatic countermeasures detailed below would put the United States in a strong position.

Most importantly, for the foreseeable future, it is exceedingly unlikely that a GBL will pose any *strategic-level* threat of the kind posed by R-spacecraft. This is because there are currently no indications in the public domain that GBLs will become capable in the foreseeable future of harming the external structures of satellites in orbits higher than LEO. At present, they are not even capable of doing so in LEO,[409] and while that is likely to change by the end of the decade[410] the lasing power supply requirements for inflicting such damage in MEO/GEO from a laser on Earth are orders of magnitude greater.[411] Additionally, the satellites in question can be readily hardened against DEW-induced heat imbalances; indeed, many—GPS included—already are.[412]

C. Jamming

(i) Threat Profile

The third threat vector selected for evaluation in this report is electronic jamming. "Jamming" refers to the deliberate use of radio signals or noise to interfere with a given target's efforts to send radio signals to, or receive radio signals from, their own satellites. Earth-based jammers may accomplish this by targeting either the *satellite* itself, known as "uplink" jamming, or the terrestrial *users* in a given area, referred to as "downlink" jamming. Jamming is already a routine feature of battlegrounds such as Ukraine and Syria, and is

---

409. "Challenges to Security in Space," DIA (2019), 20.
410. Ibid; Daniel R. Coats, "Worldwide Threat Assessment of the U.S. Intelligence Community," (Washington, DC: Office of the Director of National Intelligence (ODNI), February 2018), https://www.dni.gov/files/documents/Newsroom/Testimonies/2018-ATA---Unclassified-SSCI.pdf .
411. Ibid; For more complete coverage of the power requirements and constraints involved with GBLs, see Wright, Grego, and Gronlund, *Physics of Space Security*, 123-130, 134-135, 140-150.
412. As a result, the energy requirements and time-on-target required would be, at minimum, many tens of times greater than against a non-hardened satellite, and could easily be more; this then quickly becomes logistically infeasible, at least at technology levels comparable to the present, as it would require multiple consecutive minutes of uninterrupted high-power lasing on the same components without the target responding which is entirely unrealistic, all while consuming inordinate amounts of fuel. Carter, "Satellites and Anti-Satellites," 90.

also frequently witnessed elsewhere, including potential hotspots such as the South China Sea.

In contrast to R-spacecraft, and to some degree lasers, jamming is a threat which this report does *not* find is likely change substantially in kind over the next decade. Rather, the authors predict that we will see jamming capabilities continue to grow in prevalence, frequency, and intensity of use as they become more sophisticated, as well as increasing integration of jamming with other counterspace capabilities and operations. This is likely to occur both in military conflicts and in so-called "grey zone" contexts as China seeks to exploit the deniable and comparatively low-magnitude—but nonetheless impactful—nature of jamming in order to harry, harass, provoke, and progressively undercut the position of the U.S. and its allies in East Asia.

(ii) China's Program

China has already acquired significant jamming capabilities. U.S. intelligence assessments confirm that China is continuing, and will continue, to develop more specialized and capable jammers to counter a broader range of U.S. and allied capabilities. The Defense Intelligence Agency has specifically testified in public that,

> [T]he PLA routinely incorporates jamming and anti-jamming techniques against multiple communication, radar systems, and GPS satellite systems in exercises. China continues to develop jammers dedicated to targeting SAR [Synthetic Aperture Radar] aboard military reconnaissance platforms, including LEO [Low Earth Orbit] satellites. Additionally, China is developing jammers to target SATCOM [satellite communications] over a range of

frequency bands, including military protected extremely high frequency communications.[413]

There is also substantial evidence that the PLA is increasing both its use and operational *integration* of jamming capabilities. For example, electronic warfare (EW) capabilities were one of the many components included in the PLA's 2015 reorganization and integration of tactical and operational counterspace and offensive cyber capabilities into a Strategic Support Force (PLASSF).[414]

Despite these advances, however, there is little in the open source to suggest any major breakthroughs, either recent or forthcoming, which would alter the fundamental *nature* and unique characteristics of jamming as a counterspace threat vector.

(iii) Potential Consequences

The effects and implications of jamming are generally tactical and/or operational, rather than strategic, placing it in a very different category than R-spacecraft, and arguably even than GBLs. Nevertheless, jamming is a permanent feature of the military operations landscape in the 21st century, and can be impactful in a variety of ways. One can employ jammers to interfere with or deny access to any signal-reliant functions. Prominent use-cases for this capability include, among others: hampering communications; interfering with tracking and detection; and denying access to the PNT systems (e.g., GPS) that are increasingly central to both military and civil/commercial positioning and weapons targeting in the modern era.

---

413. "Challenges to Security in Space," DIA (2019), 20.
414. Kania and Costello, "Seizing the commanding heights."

DIA has testified publicly that China's jamming is often "directed at ground [and airborne] users such as a ground forces unit [or airplanes] using satellite navigation to determine their location."[415] This can complicate battlefield operations and tactics. If not adequately prepared-for and countered, jamming risks undercutting the ability of the United States to reliably operate in a given area, especially when projecting power over the horizon; the American way of war has become increasingly reliant on PNT capabilities in recent decades, a fact of which the PLA is well aware. Even short of actual operational impact, perceptions of jamming effectiveness could undercut the confidence of the U.S.'s allies and partners in its ability to operate freely, or run the risk of emboldening adversaries.

Another concern—and, given the various countermeasures readily available to the U.S. military, arguably a greater one—is the use of jamming to harass civilian and commercial targets. By its very nature, jamming is reversible: once the jamming signal is turned off, the targeted services can resume.[416] Jamming is also relatively deniable: DIA has noted publicly that electronic jamming can be "challenging to attribute and distinguish from unintentional interference."[417] These features make it attractive for so-called "grey zone" use—employment in operations which are well short of armed conflict but are nonetheless hostile. If not adequately countered, this creates significant potential for direct or indirect harm to U.S. allies, and the international rule of law. For example, one could easily imagine China attempting to blackmail Japan or Taiwan in a crisis by using jammers to interfere with their ability to land commercial airliners safely and reliably.

---

415. DIA, "Challenges to Security in Space," 9.
416. Pavel Velkovsky, Janani Mohan, and Maxwell Simon, "Satellite Jamming: A Technology Primer," *On the Radar*, Center for Strategic and International Studies, April 3, 2019, https://ontheradar.csis.org/issue-briefs/satellite-jamming.
417. DIA, "Challenges to Security in Space," 9.

Fortunately, all of these use-cases for jamming can be readily mitigated, countered, and responded to with the proper combination of unilateral and diplomatic measures. Those measures are detailed *infra*, at Section III(B)(iii).

III. Countering These ASAT Threats: Diplomatic & Unilateral Measures

The United States must move quickly if it is to maintain stability and prosperity in orbit through the end of the 2020s and beyond. In order to do so, the U.S. will have to transcend traditional divisions between "hawks" and "doves," and adopt a nuanced combination of both unilateral and diplomatic measures. Unilateral and multilateral approaches are mutually-reinforcing, and neither will succeed without the other. The U.S. will need to carefully align law, doctrine, and policy; make full use of all instruments of national power, including diplomacy, economic measures, and the prospect of military action; and coordinate extensively with international stakeholders, including both allies and potential adversaries.

The three ASAT threat vectors evaluated in this report are no exception. To minimize the risk of strategic surprise and ensure that all necessary ASAT countermeasures are successfully identified, this report proposes a two-tiered methodology for counterspace solution identification. First, a set of one-size-fits-all measures should be adopted, establishing a strong foundation for mitigation, deterrence, and countering of counterspace threats generically. The majority of thinking and action by the United States on counterspace countermeasures to date falls into this category. Substantial progress has been made on these measures, especially over the past ten years, yet serious shortcomings remain.

In the past, the assessment has often ended with adoption of these generic measures. There has been little effort to systematically identify what current and future threat vectors might escape coverage by these measures, much less to design and adopt appropriate tailored countermeasures. This is especially concerning, because not only does the U.S. critically rely on space for a wide spectrum of civil, commercial, and military functions, but the domain is increasingly offense-dominant. Therefore, failure to prepare adequately for even a single threat vector, if it has sufficient operational and/or strategic impact, has the potential to produce severe economic and political costs, or risk starting or escalating a large-scale conflict. At the same time, not all threat vectors are created equal, and to treat them as though they all threaten the same degree or type of impact is to guarantee we will fail to adequately account for the specific characteristics of each threat vector.

Therefore, this report proposes a crucial second analytic step, in which these generalized measures are treated as an assumed *baseline* against which specific threat vectors must be further evaluated on a case-by-case basis. Doing so yields a second essential category of ASAT countermeasures: supplemental measures specifically tailored to fill any gaps in these generic protections that might be exploited or even created by the unique characteristics of a specific ASAT threat vector.

While such careful case-by-case assessment may have always been a best-practice, the U.S. has, to date, been able to largely skate by without it. This is primarily because the landscape of presently-existing counterspace threats has remained more or less stagnant for decades. Now, as a range of emerging technologies and the accelerating proliferation of counterspace capabilities combine to produce ASAT threats fundamentally different in nature from those that the U.S. has faced for the past half-century, specificity is no longer a luxury, but a strategic necessity.

Thus, Section III proceeds with a synergistic approach consisting of two parts. First, Part A briefly surveys some of the most important measures of general applicability for the U.S. to adopt and the reasons that these measures are necessary. Second, Part B examines each of the three ASAT threat vectors in turn, in accordance with their threat profiles and potential consequences as set forth *supra* at Section II, and identifies the set of supplemental tailored measures that the U.S. should adopt to effectively counter each one.

D. Measures of General Applicability

The first layer of the U.S.'s defense against counterspace threats should be a suite of measures of general applicability. These diplomatic and unilateral measures may be usefully divided into two categories—measures focused on *denial* and measures focused on *response*.

(i) Generic Denial Measures

It is imperative for the United States to minimize, wherever possible, the expected utility of an adversary's hypothetical ASAT operations. In so doing, the U.S. may not only minimize the direct consequences should the worst-case come to pass, but may also alter adversaries' calculations regarding the value of conducting such attacks (or perhaps even acquiring such capabilities) in the first place. Limiting the severity of potential ASAT attacks will also serve to increase U.S. freedom of action by reducing the probability of irresistible pressures to escalate either preemptively or in retaliation.

The first and most important defensive measure is distributing capabilities. Wherever possible, the United States should replace or augment current U.S., partner, and commercial capabilities with capabilities from proliferated constellations of more, smaller, less expensive satellites. This is most feasible for communications and environmental monitoring. Other missions such as PNT, signals intelligence (SIGINT), and missile warning may also benefit from these approaches. However, other missions, such as those that require large mirrors for high resolution imagery or high power for radar imagery are unlikely to be suitable for distributed architectures—at minimum during the timeframe under consideration (i.e., the remaining of this decade), and possibly for much longer.

Second, the U.S. should continue to improve its capability to rapidly reconstitute key assets, especially via operationally-responsive rapid-launch capability into LEO. These reconstitution efforts should especially prioritize any assets which, while being low-orbit and relatively replaceable, are not amenable to distribution.

Third, the United States should prioritize the fielding of multiple redundant systems for all essential strategic and operational functions (e.g., theater-wide comms). Furthermore, the authors strongly advise that a hypothetical invasion of Taiwan with early counterspace attacks serve as one primary referent for contingency planning. In particular, the U.S. should ensure not only acquisition but also forward pre-positioning of airborne backups wherever possible, ideally themselves distributed and capable of functioning from standoff range.

Fourth, for critical systems where distribution or rapid reconstitution are impractical (GPS and the NC3 assets being two such examples discussed in this report), U.S. critical satellites should have built-in countermeasures and/or bodyguard spacecraft for protection against the full spectrum of known and predictable counterspace threats. Where possible, the U.S. should also seek to design-in sufficient modularity to enable future *in-situ* upgrading via satellite servicing as threats and countermeasures advance without the need for a lapse in coverage or loss in service.

Fifth, where possible, the United States should work with key allies to enable U.S. military use of their space assets in the event that our own satellites are destroyed and disabled, and vice-versa. At minimum, the U.S. should prioritize equipping U.S. military devices with the ability—already common in civilian electronics—to flexibly make use of allies' PNT constellations (e.g., GALILEO, QZSS) in the event of GPS degradation or disruption; and should increase commingling and 'piggybacking' of U.S. and allied satellites in an effort to complicate adversary targeting decisions. Additionally, the U.S. should employ simulations and exercises with stressing scenarios to build partner capacity and improve interoperable capabilities.

Sixth, the U.S. should renew[418] and expand ongoing space security dialogue—and associated bilateral channel(s)—with China. These should be designed to serve two critical functions: first, to build high-level peer-to-peer networks, establish communication channels for use in a crisis, and ideally facilitate increased exchange of SSA data, all of which may play important roles in reducing the risk of misunderstanding or misperception; and second, to inculcate mutual understanding, and to enable the U.S. to clarify and reiterate on an ongoing basis the scope and seriousness of its redlines

---

418. Craig Kafura, "Renew Space Dialogue with China," War on the Rocks, February 10, 2022, https://warontherocks.com/2022/02/renew-space-dialogue-with-china/.

regarding critical space-based assets, especially those with acknowledged strategic nuclear roles.

(ii) Generic Response Measures

To credibly deter or respond effectively to ASAT attacks, or the threat of such attacks, the United States must urgently develop, adopt, and publicly clarify its ability to swiftly impose clear and *proportional* costs. The U.S.'s recent experience in Ukraine provides a clear positive example of the tremendous value and importance of such preparedness.

To this end, the U.S. should first and foremost orient planning and processes around development and implementation of an increasingly integrated cross-domain approach to deterrence and cost-imposition. The authors recommend that this includes development of a more robust suite of counterspace capabilities. However, whether or not it acquires such capabilities, the United States government must stand ready at the bureaucratic Inter-agency level to implement a wide array of other proportionate graduated-response options depending on the situation, and must clearly communicate this to adversaries with some degree of specificity. Such measures might include, among many others: conditionality of access to certain export markets; light and heavy sanctions on CCP officials or key economic sectors or Chinese banks; a stronger stance on territorial disputes (e.g. increasing forward deployment around the Senkaku Islands, or even recognizing Japanese sovereignty over the Islands); limited strikes conducted against Chinese infrastructure in the South China Sea; damaging or disabling cyberattacks conducted against valuable targets in China; increased deployment of missile defense capabilities in and around the Korean Peninsula and Japan; increased provision of long-range strike capabilities to Japan and South Korea; and a wide range of other levers.

Second, the authors recommend that the U.S. prioritize comprehensive mapping of Chinese space situational awareness (SSA), cyber, and command and control (C2) capabilities supporting the PLA's kill chain for attacking U.S., partner, and commercial space-based national security assets. The U.S. should prepare graduated non-kinetic and kinetic options to disrupt key nodes in this kill-chain, some of which may be outside mainland China and may be more susceptible to diplomatic pressure or present less hardened targets.

Third, the United States must consult closely with its allies if these measures are to be effective. Close coordination, especially with treaty allies, is not only important for its own sake, but will help to ensure that the alliance is prepared to act swiftly and in lockstep in the event of either direct ASAT attacks or attempted 'grey zone' coercion. Crucially, the U.S. should seek specific consensus with its allies and partners regarding the legality and desirability of various countermeasures and uses of force under a variety of different foreseeable scenarios. To the extent possible, the U.S. and its allies must seek to come to common understandings regarding key issues such as redlines, warning/self-defense zones, acceptable preemptive self-defense measures, and so on.

While some progress has been made on all of the measures identified above, every single one of them requires increased prioritization and investment. In some cases, efforts are maturing but not yet complete. In far more cases, efforts are still in the nascent stages. Given how rapidly adversary capabilities are maturing, this is highly alarming. This makes clear why officials at the most senior level of government have been sounding the alarm.[419] Assuming that the

419. To cite one example among many, Gen. John Hyten noted shortly before his retirement that that "the DoD is still unbelievably bureaucratic and slow" in its response to China's rapidly advancing space weapons. See: Sandra Erwin, "Hyten blasts 'unbelievably' slow DoD bureaucracy as China advances space

U.S. moves swiftly to rectify this and accelerate implementation of this set of countermeasures, however, it will have established a strong generalized foundation for defense against counterspace threats.

E. Supplemental Measures Tailored to the Three Threats

(i) Measures to Counter Rendezvous Spacecraft

A central conclusion of this report is that while one-size-fits-all measures are certainly a necessary component of the defense against counterspace threats, they are not sufficient on their own to counter the risks produced by next-generation ASAT technologies. Careful attention to the unique profile of each threat vector reveals that some ASATs are likely to require supplemental tailored measures in order to be successfully deterred or countered. The first example of this, and in the view of the authors by far the most urgent at the present time, is R-spacecraft.

As illustrated in Section II, the threats posed by R-spacecraft fall into two distinct categories. First, R-spacecraft pose a unique threat to vital satellites in GEO, HEO, and MEO with the potential for severe strategic- and operational-level impacts, particularly by the use of pre-positioning in close proximity to key strategic assets (e.g., nuclear command-and-control assets in GEO and HEO, and GPS satellites in MEO). Second, R-spacecraft may be employed in threatening or unsafe ways in LEO, or be perceived as such, with potential consequences for civil and commercial space traffic, tactically-significant military and intelligence assets, and the space assets of U.S. allies and partners. These are distinct threat vectors and require two distinct solution sets.

---

weapons," SpaceNews, October 28, 2021, https://spacenews.com/hyten-blasts-unbelievably-slow-dod-bureaucracy-as-china-advances-space-weapons .

Countering the Threat to Critical Assets in GEO, HEO and MEO

Pairing increased redundancy and resiliency measures with an expanded array of flexible graduated response options is an effective formula for addressing most counterspace threats. However, many of the U.S.'s most strategically vital satellites will remain vulnerable to R-spacecraft attacks for at least the next decade and likely much longer, even if planned upgrades are accelerated. Examples include the 6 GEO satellites and 4 HEO hosted payloads in the Space-Based Infrared Systems (SBIRS)[420] which together constitute the U.S.'s primary space architecture for global missile launch detection and warning, the 6 GEO satellites in the Advanced Extremely High Frequency (AEHF) Satellite System[421] which provide tactical communications for conventional forces and strategic communications for nuclear forces, and the GPS satellites discussed above, among others.[422] These constellations exemplify the paradigm of exquisite systems: each consists of a small number of large and very expensive (about $1.7 billion each) satellites.[423] This is largely unavoidable at current levels of technology, because there

---

420. "Space-Based Infrared System (SBIRS)," Aerospace Technology, https://www.aerospace-technology.com/projects/sbirs/ .
421. "Advanced Extremely High Frequency (AEHF) Satellite System," Air Force Technology, March 19, 2021, https://www.airforce-technology.com/projects/advanced-extremely-high-frequency-aehf/#:~:text=The%20AEHF%20system%20consists%20of,for%2024%20hours%20a%20day
422. "Nuclear Command, Control, and Communications (NC3) Modernization," Congressional Research Service, December 8, 2020, https://sgp.fas.org/crs/nuke/IF11697.pdf .
423. Brian Chow, "Op-ed: Nuclear vulnerability: In-orbit bodyguards would help protect NC3 satellites from attack," (SpaceNews, April 2019), https://spacenews.com/op-ed-nuclear-vulnerability-in-orbit-bodyguards-would-help-protect-nc3-satellites-from-attacks/; Brandon Kelley and Brian Chow, "Countering Chinese Rendezvous Spacecraft, Ground-Based Lasers, and Jammers in the 2020s and Beyond," Presentation in advance of wargame (Washington, DC: Nonproliferation Policy Education Center (NPEC), April 2021).

are some capabilities which cannot be adequately substituted for by smaller distributed constellations or provided at lower cost.[424] In contrast, an R-spacecraft attacker is likely to be cheaper than such satellites by one to two orders of magnitude because, once the RPO software and robotic-manipulation software have been developed and sufficiently tested and appropriate hardware component manufacturing and assembly lines established, these small satellites can be relatively cheaply and quickly mass-produced. To disincentivize such attacks, as well as successfully counter them should they occur, the U.S. will require countermeasures comparable in cost to these attackers.

Some of these legacy constellations are being replaced, but follow-on satellites will remain similarly expensive and vulnerable to R-spacecraft. For example, the follow-on to the SBIRS will be the Next-Generation Overhead Persistent Infrared Program (Next-Gen OPIR), which will consist of 3 GEO satellites slated for launch in 2025 and 2 HEO satellites in 2028.[425] This is in line with the planned timeline to have all five satellites in operation beginning 2029, with Next-Gen OPIR expected to be in service well into the 2030s or

424. As NRO director Christopher Scolese put it, the U.S. will continue to maintain "some number of large satellites to address questions that *only* they can address." Sandra Erwin, "Scolese: NRO advancing space technology, developing tactics to defend satellites," (SpaceNews, 2019), https://spacenews.com/scolese-nro-advancing-space-technology-developing-tactics-to-defend-satellites/ .

425. "Next-Generation Overhead Persistent Infrared Program Selects Mission Payload Suppliers," Space Systems Command, March 1, 2022, https://www.ssc.spaceforce.mil/Portals/3/Documents/PRESS%20RELEASES/Next-Generation%20Overhead%20Persistent%20Infrared%20Program%20Selects%20Mission%20Payload%20Suppliers%20v4.pdf?ver=7qUXGXb35DOGSCh2uAQEGw%3D%3D .

beyond.[426] These five satellites cost over $1 billion each.[427] Similarly, the follow-on to the AEHF satellite system is the classified Evolved Strategic Satcom (ESS) program. Limited public information indicates that there will be two GEO constellations separately serving AEHF satellite system's two missions of tactical and strategic communications, and that the total cost will be over $8 billion from 2023 through 2027.[428] Therefore, it seems that the ESS program will retain the vulnerability of its predecessor, and its satellites will continue to need bodyguard or other similar protection, again during their services well into the 2030s.

Finally, even if some medium-term technical solution could be found, the threat outlined herein will mature well before it could be built and deployed,[429] so finding an interim solution would remain paramount. Indeed, a time-limited window of vulnerability might well be the *worst possible* scenario, because it may uniquely incentivize exploitation of the vulnerability before the opportunity slips away. This is a particularly salient concern given that many of the likeliest contexts for such pre-positioning to occur are ones in

---

426. Theresa Hitchens, "SMC, SDA Missile Warning Sat Ground Systems Won't 'Talk' – For Now," (Breaking Defense, January 2021), https://breakingdefense.com/2021/01/smc-sda-missile-warning-sat-ground-systems-wont-talk/.
427. Brian Everstine, "Lockheed Receives Up to $4.9 Billion for Next-Gen OPIR Satellites," Air Force Magazine, January 5, 2021, https://www.airforcemag.com/lockheed-receives-up-to-4-9-billion-for-next-gen-opir-satellites/ .
428. Sandra Erwin, "U.S. to ramp up spending on classified communications satellites," SpaceNews, May 1, 2022, https://spacenews.com/u-s-to-ramp-up-spending-on-classified-communications-satellites/#:~:text=U.S.%20to%20ramp%20up%20spending%20on%20classified%20communications%20satellites,-by%20Sandra%20Erwin&text=WASHINGTON%20%E2%80%93%20The%20U.S.%20Space%20Force,according%20to%20DoD%20budget%20documents.
429. See, e.g., generally: Chow, "Space Traffic Management"; Kelley and Chow, "Countering Chinese Rendezvous Spacecraft"; Chow, "Stalkers in Space"; *supra* at Section II discussion of the emerging capabilities, as well as current and near-future developments.

which a potential adversary seeks to undertake a high-value territorial incursion while deterring U.S. response, in an effort to undercut the credibility of the U.S. and the strength of its global system of alliances.[430]

Similarly, reconstitution of GPS would be prohibitively expensive. Successfully denying GPS access to an entire area for a prolonged period would also likely have serious ramifications even if later rectified; even temporary denial may, for example, suffice to facilitate an attempted PLA seizure of Taiwan. Additionally, as discussed in Section II and the Appendix, China may have on the order of 200 ASAT R-spacecraft in the near future, perhaps as soon as 2026. They can disable most, if not all of the three dozen or so GPS satellites with plenty left to attack other critical satellites. Again, as discussed in Section II, the impact in aiding Taiwan and fighting a war with China will be serious in our loss of lives and treasure, if not losing the war.[431] The U.S. is working to field non-space-based backup PNT systems and increase the ability to easily switch to signals from allied PNT systems (e.g., GALILEO and QZSS). Such efforts should continue, and are likely to be somewhat effective at mitigating the tactical and operational consequences of GPS loss or interference. It is possible, but uncertain, that this could lead potential adversaries to avoid targeting GPS satellites in the

---

430. For quality assessment of the 'credibility thesis,' see Hal Brands, Eric Edelman, and Thomas G. Mahnken, "Credibility Matters: Strengthening American Deterrence in an Age of Geopolitical Turmoil," (Washington, DC: Center for Strategic & Budgetary Assessments (CSBA), 2018), https://csbaonline.org/uploads/documents/Credibility_Paper_FINAL_format.pdf. The broader strategic deterrence and crisis stability challenge produced by R-spacecraft is in some respects reminiscent of the "Senkaku Paradox" described by Michael E. O'Hanlon in the context of probing territorial redlines; see Michael O'Hanlon, *The Senkaku Paradox: Risking Great Power War Over Small Stakes* (Washington, DC: The Brookings Institution, 2019).
431. Brian Chow and Brandon Kelley, "China's Anti-Satellite Weapons Could Conquer Taiwan—Or Start a War," The National Interest, August 21, 2021, https://nationalinterest.org/feature/china%E2%80%99s-anti-satellite-weapons-could-conquer-taiwan%E2%80%94or-start-war-192135.

first place by making them unattractive targets. Nevertheless, it is prudent to ready an effective response in case they are not so deterred.

Clearly, therefore, it is necessary to find some means of limiting the R-spacecraft threat. But how might this be accomplished? As mentioned in Section II, a capability-focused control measure is wholly infeasible, and would be undesirable in any case. The authors strongly advise that the United States adopt a set of specialized legal and policy measures, on both a unilateral and multilateral basis, which are directly tailored to *limiting the ability to pre-position* R-spacecraft in close proximity to high-sensitivity targets. While little can be done to outright prevent pre-positioning far from a target, and/or the initiation of a swift attack from far away, the goal should be to ensure sufficient *warning* and visibility at a minimum, and ideally to provide means of countering the strike. [432] As Carter puts it, the concerning scenario is an attack which is "neither time-consuming *nor* obvious [:] If the attack takes time, surprise is lost. If the source of the attack is obvious, the attack itself constitutes warning."[433] Exhibit A: "Even a space mine lurking one degree away in the [GEO] arc…and possessing substantial propellant would still take several minutes to arrive at its quarry. During this time on-board sensors or a distant space track[ing] system could give warning…With adequate attention to…survivability…and some ability to keep space mines at a distance, therefore, the U.S. might have little to fear…."[434] However: in contrast to R-spacecraft placed either, like this one, "in a distant part of the same orbit, which would require it to maneuver to approach and attack the target," or in "a crossing orbit in the same…or … a different [orbital] plane,"

---

432. See, e.g., the discussion of ASATs in "crossing orbits"—contrasted with "trailing orbits," of which the GEO and MEO discussions here are a subset—in Wright, Grego, and Gronlund, *Physics of Space Security*. Occurs throughout, including e.g. at 157.
433. Emphasis added. Carter, "Satellites and Anti-Satellites," 79-80.
434. Ibid.

an ASAT which is "co-orbital with the target satellite...keeping within a fixed distance...could attack almost instantaneously[.]"[435] As Grego et al. note, the physics involved mean that, in the latter case, one can attack "on the scale of minutes," whereas "an ASAT in a crossing orbit may require hours before it is in position to attack."[436]

This, of course, raises the question of what specific limits should be pursued and enforced, and by what means. The authors recommend that the U.S. adopt three key measures.

First, the United States must formulate a clear framework for application, and pursue international recognition—agreement, if possible, awareness if not—of this framework. For over 35 years, authors have proposed that various types of "zones" be established to regulate on-orbit activity—e.g., self-defense zones,[437] keep-out spheres,[438] safety zones,[439] and, most recently, warning zones.[440] These proposals vary widely in their provenance and their details. What they all share in common, however, is an understanding of the need to shield our most critical but vulnerable satellites from harm by deterring or countering unauthorized close approaches under at least certain conditions.

---

435. Wright, Grego, and Gronlund, *Physics of Space Security*, 152-154.
436. Wright, Grego, and Gronlund, *Physics of Space Security*, 156.
437. Chow, "Stalkers in Space."
438. "CONFERS Recommended Design and Operational Practices," Consortium for Execution of Rendezvous and Servicing Operations, February 1, 2019, https://www.satelliteconfers.org/wp-content/uploads/2019/02/CONFERS-Operating-Practices-Approved-1-Feb-2019-003.pdf.
439. CONFERS "CONFERS Recommended Design."
440. Brandon Kelley, David Bernstein, Michael Cerny, and Raphael Piliero, "Countering Co-Orbital ASATs: Warning Zones in GEO as a Lawful Trigger for Self-Defense," (Washington, DC: Nonproliferation Policy Education Center (NPEC), June 2020), Reprinted as Ch. 5 in *Space and Missile Wars: What Awaits,* Henry Sokolski, ed., (NPEC, April 2021), http://npolicy.org/article_file/Space_and_Missile_Wars-chapter_5.pdf.

Comprehensive discussion of these zones, their design, and relative merits of different proposals is outside the scope of this report. However, the authors recommend that, at a minimum, the U.S. should publicly clarify—jointly with allies if possible, or unilaterally if not[441]—that there exists some point at which R-spacecraft maneuvering in sufficiently close proximity to critical target satellites would rise to the threshold that the zone defender would have the right to interrogate the invader and prevent it from causing harm to the satellite(s) in the zone, perhaps through the use of force if necessary. These acts of interrogation and self-defense can and should be designed to be non-escalatory, including in some cases even causing little permanent damage to the invaders.

Doing so would be clearly and definitively lawful, as it would clarify only that there exist certain conditions under which the behavior of a given spacecraft rises to the threshold of constituting evidence that an armed attack can be "imminent," and preemptive self-defense is necessary. Put differently, the goal is to clarify what constitutes *sufficient evidence of an imminent threat* and under what conditions force might be the *only means of reliably forestalling that threat*. In some cases, the resultant "force" may be as gentle as bringing the invader out of the zone without permanently damaging the invader. In others, it might involve disabling the offending asset or retaliating proportionately across domains.

On the other hand, it is unclear under what conditions those requirements might be satisfied in LEO, due to the orbital parameters involved, the generally less-critical nature of the

441. For one illustration of this synergistic and multi-layered approach in action, as well as a clear demonstration of why initiating it now is necessary, refer to after-action report for the 2021 China space wargame sponsored and hosted by NPEC: "China Waging War in Space: An After-Action Report," Edited by Henry D. Sokolski, Nonproliferation Policy Education Center, August 2021, https://npolicy.org/wp-content/uploads/2021/08/2104-China-Space-Wargame-Report.pdf .

satellites at risk, and the availability of other effective solutions (e.g., maneuvering out of the way, using reversible measures such as dazzling/lasing using GBLs or using jamming/EW or cyber capabilities, etc.). However, this is not the case for NC3 satellites in GEO and HEO, nor for PNT assets in MEO. Once R-spacecraft are positioned within sufficiently close to these satellites, China would be far more confident of success in initiating hostile RPOs, with no warning to enable defense. Nor can the blow be confidently absorbed by us: there are only a small number of them capable of performing the necessary missions, and they are acutely vulnerable. The only means of effective defense is to prevent such prepositioning in the first place.

Second, the U.S. must urgently move to acquire, deploy, and publicly disclose the existence of capabilities for verification and enforcement. For this purpose, the U.S. should field and deploy bodyguard satellite(s) around every critical satellite.[442] A critical satellite is defined as one that is vulnerable without the protection from bodyguard and that we do not want to fight a war without this satellite. Should the above thresholds ever be crossed, it is these bodyguards which would in most cases be charged with interrogation and self-defense against the invaded spacecraft.[443] They could also be useful for space domain awareness purposes, particularly in characterizing the space environment in the

---

442. For more regarding the use of bodyguards for monitoring and enforcement of zones, see: Chow, "Op-ed: Nuclear vulnerability: In-orbit bodyguards"; see also the other Chow pieces cited above. For more on the cost-effective leveraging of cubesats for active defense of satellites, see Michael Nayak, "Deterring Aggressive Space Actions with Cube Satellite Proximity Operations: A New Frontier in Defensive Space Control," *Air and Space Power Journal* vol. 31 no. 4, 92-103, https://www.airuniversity.af.edu/Portals/10/ASPJ/journals/Volume-31_Issue-4/SEW-Nayak.pdf.

443. Ibid.

immediate vicinity of their wards.[444] Realistically, the fastest type of bodyguard spacecraft to develop in the next few years would be specialized R-spacecraft, which because of their narrower and less complicated role can be optimized to be much smaller and cheaper, reversing the cost-effectiveness advantage. For the longer term, the U.S. should seek to equip bodyguards with additional reversible measures, such as jamming, laser, and cyber to prevent the invaders from reaching our satellites in the zones. Additionally, such bodyguards in principle may also help to counter other ASATs, such as DA-ASATs as onboard countermeasures evolve and miniaturization continues. Ironically, while U.S. efforts to address satellite vulnerability were triggered by China's successful test of DA-ASAT in 2007, the U.S. has yet to even propose a means of defending NC3 satellites and GPS against DA-ASATs.[445] With time fast running out as these threats mature, the U.S. should move urgently to begin mass-production and deployment of such bodyguard spacecraft. Now is the time for critics and naysayers to

---

444. Ibid. Even critics and skeptics concede that some version of bodyguards would be quite useful for some sorts of applications—for example, Weeden: [...] I think guardian satellites are a lot less useful in practice than often talked about. And again, this has to do with orbital mechanics. They can be used to increase awareness in the direct area around the principal satellite. They can...complicate an adversary's RPO if they're trying to approach another satellite.... And I think they could be a platform for certain types of countermeasures. But aside from those aspects, guardian satellites are pretty limited compared to how they're often described in the media.
Brian G. Chow and Brian Weeden, "Transcript of a Discussion." There are real (albeit decisively overcomeable) concerns to be raised regarding feasibility/viability, potential to stoke weaponization fears, etc.; but ultimately, most of the solution is not to avoid bodyguards but rather to be clear-eyed about what we build them to do and what roles we task them with implementing. Most importantly, how they can be used for interrogation and self-defense without causing escalation.

445. Though, for a range of reasons alluded to above and further explored elsewhere, DA-ASATs are substantially less likely to be used for NC3- or GPS-attacked mission than R-spacecraft.

openly outline their concerns to enable collaboration to either find solutions to them or identify a preferable alternative.

Third, the U.S., either alone or in concert with its allies and partners, should carefully consider whether there exist cases in which the use of preemptive responses *other* than direct prevention of the attack would be both lawful and preferable, and, if so, what sorts of measures and under what conditions. The results of these deliberations must be codified and integrated into U.S. doctrine, and the U.S. absolutely must follow through. Additionally, the results should be publicly communicated with a level of detail and specificity for them to serve their desired deterrent functions. Finally, the international legal rationales for these conclusions should be publicly articulated in full.

Further work should be done to comprehensively evaluate all of these details, and the authors intend to publish further on the subject in the near future.

Countering the Persistent Threat to Satellites in LEO

The LEO threat is treated here only briefly, for two key reasons. First, unlike in higher orbits, many of the greatest *counterspace* threats that RPOs might otherwise pose in LEO may be heavily mitigated via the proper combination of one-size-fits-all defensive measures and deterrent capabilities outlined above. One possible exception may be non-distributable and high-cost LEO satellites such as legacy systems providing very high-resolution imaging capabilities, which may be vulnerable and attractive targets for R-spacecraft attacks, in which case a modified variant of the zones-and-bodyguards proposal may be warranted for these satellites. Second, the LEO threat is best understood and dealt with as an *STM* challenge. Critically, STM should provide an effective framework and incentive structure for limiting both accidental collisions between space objects *and* intentional disablement of satellites. The potential for 'grey zone' coercion by means of threats or low-level interference with the assets of civil agencies, companies, or U.S. allies is also a serious concern.[446] However, to successfully drive the timely development of an appropriate framework and adequately incent comprehensive adoption and compliance, the United States must urgently pursue two key avenues, each of which will require an associated set of unilateral and diplomatic measures. The authors intend to publish a separate paper dealing with this in detail in the near future, so it is outlined here only briefly. First, independent of negotiations regarding STM or rules of the road, the United States should work with its allies to elaborate interpretations of existing international law, especially the Outer Space Treaty (OST), that would impose meaningful limitations on these harmful behaviors and help to regulate peaceful activities in LEO. The authors strongly advise that particular emphasis be placed on OST Art. IX,[447] the provision which requires that states act with due regard and avoid harmful interference. In order to give these provisions teeth, the U.S. must determine and clarify what sorts of countermeasures it understands to constitute lawful means of restoring compliance in the event of a significant breach.

Second, recognizing that China is highly unlikely to accept our desired norms of space behavior and will almost certainly continue to weaponize international fora to frustrate consensus, the United

---

446. This was, for example, a major part of the scenario contemplated by NPEC's China space wargame; see Sokolsky, NPEC, "China Waging War in Space."
447. "Treaty on Principles Governing the Activities of States in the Exploration and Use of Outer Space, including the Moon and Other Celestial Bodies," United Nations Office for Outer Space Affairs, https://www.unoosa.org/oosa/en/ourwork/spacelaw/treaties/outerspacetreaty.html .

States should adopt a dual-track approach,[448] neither disengaging from international institutions nor allowing lack of consensus to prevent it from working unilaterally and with allies to further its preferred norms and interpretations. These norms and interpretations should, at root, be based on a just and fair principle: no country should be able to take advantage of any STM agreement to threaten another country's critical satellites during peacetime or in a crisis. Specifically, in addition to leading all other countries to negotiate an international STM regime, the U.S. must work both unilaterally and on a bilateral and multilateral basis with its allies and partners to establish a Western STM regime. Access to the lucrative Western space markets should be made conditional on adherence to an STM regime, whether Western or international, including explicit rules limiting pre-positioning in threateningly close proximity to another country's critical satellites without prior consent. In this manner, the U.S. and its allies might be able to garner sufficient leverage to incentivize other countries, including most notably China and Russia, to consent to these internationally agreed-upon rules of the road and remain in strict compliance with them. Should China or Russia refuse to join either STM regime, or subsequently cheat or defect, the Western critical satellites would remain protected by the rules in the Western STM, and there would be clear and significant costs for their intransigence. Limiting other countries' ability to threaten these satellites (thus dramatically reducing the benefits of non-compliance) while simultaneously pegging continued participation in the lucrative Western space market to adherence to an STM regime (thus substantially increasing the

448. Brian G. Chow, "Op-ed | How to convince China and Russia to join a space traffic management program for peace and prosperity," SpaceNews, January 26, 2021, https://spacenews.com/op-ed-how-to-convince-china-and-russia-to-join-a-space-traffic-management-program-for-peace-and-prosperity/.

benefits of compliance) will provide a powerful incentive for countries to be 'good citizens' in orbit.[449]

(ii) Measures to Counter Lasers

In contrast to pre-positioning of R-spacecraft, the core suite of generally-applicable measures would, if it were properly implemented, probably be sufficient to counter the majority of the threat that lasers are likely to pose over the next five to ten years. There are reasons for this. First, the potential impact of laser attacks may be effectively mitigated for new space systems through a combination of proliferated constellations for functions which are capable of being distributed, and hardening for exquisite systems that cannot be. This limits the potential impact of laser attacks and, because of the heavily reduced expected payoff, should rationally also limit the likelihood that they will be attempted in the first place. Second, the impact of even successful lasing will tend to be relatively small-scale and tactical in nature. Therefore, unlike in the context of nuclear early warning satellites, for example, there are no acute first-use pressures at play for either party. Third, the construction, testing, and use of lasers is both highly visible from orbit and immediately attributable, rendering it easy to either avoid or preempt. (Note, however, that as mobile lasers become more powerful, this may complicate both avoidance and preemption.) Most importantly, it is relatively easy, with appropriate preparations, for the U.S. to respond credibly and proportionately to laser attacks in a manner of the U.S.'s choosing. This provides a clear mechanism for deterring laser attacks, or responding should they occur. Nevertheless, this report does find that a number of tailored diplomatic and unilateral measures are likely to be helpful for ensuring that these general measures will be effective in this context,

449. Ibid.

as well as to further attenuate the potential impact, both now and especially beyond the end of the decade.

First, the United States should unilaterally establish standard operating procedures (SOPs) limiting the impact of sensor damage by ground-based lasers.[450] Specifically, the U.S. should mandate and routinize the use of the lowest-resolution imagery sufficient for a given task, increase integration of commercially-available imagery, and routinize imaging during peacetime as the chances of damage are much higher during a crisis. As a result, our limited number of valuable high-resolution imaging satellites will be less often exposed to potential lasing.

Second, the U.S. should engage its allies diplomatically, seeking to integrate persistent lower resolution imagery into allied kill chains and tip and cue higher fidelity allied assets.

Third, the U.S. should seek international agreement limiting such lasers. A strong agreement *prohibiting* such lasers is overwhelmingly unlikely: they are extremely attractive to China as a means of countering prospective U.S. space-based ballistic missile defense (BMD), as well as achieving operationally significant effects on significant U.S. intelligence and military capabilities, and of conducting hostile 'grey zone' activities against the civil and commercial assets of the U.S. or its allies. Moreover, even if such a deal were plausible, there would be a strong likelihood of cheating: China can always plausibly assert that high-power lasers are necessary and intended solely for some civil and peaceful purpose, such as long-range asteroid detection or highly accurate satellite and space debris tracking in the service of the global community. Instead, the U.S. should seek several agreements. First, the U.S. should seek Chinese agreement to mutually prohibit harmful interference with national technical means, akin to the longstanding

450. Chow and Sokolski, "U.S. satellites increasingly vulnerable."

agreements between the U.S. and Russia. They may well not agree to such a measure, but the U.S. should make the effort. It may succeed, but even a failed attempt to broker a deal would provide valuable information to the international community regarding China's intent and motives, as most countries would turn an appropriately skeptical eye on their opposition to treaty verification. Second, because China's ground-based lasers may go mobile sometime in the latter half of the 2020s or beyond, the United States and its allies should minimize the risk that mobile lasers will be used as ASATs by pushing for international agreement to register all mobile lasers (including their locations) that are usable or convertible for ASAT purposes. Such an agreement would optimally require one month's advance warning of intended movement and real-time broadcasting of said movement. Compliance with such an agreement, if one were reached, could be adequately tracked by the public via commercially-available imagery, and separately on a classified basis by U.S. national technical means (NTM). Regardless of whether such an agreement is reached, the U.S. intelligence community should be tasked with tracking locations of mobile lasers.

Fourth and finally, the United States must remain ready to implement whatever unilateral measures might be necessary to deter, counter, disable, or prevent or respond to the use of China's GBL stations. In the most extreme circumstances, these bases can be readily located and are highly vulnerable to aerial attacks, making strikes at the source a potential response to this threat; however, in nearly all conceivable circumstances there are likely to be much more proportionate and less escalatory deterrent and response options available, particularly if the U.S. fully implements the suite of general-purpose recommendations outlined in Part A.

(iii) Measures to Counter Jamming

Finally, jamming. As with lasing, the lower stakes of successful jamming place it in a different category than a threat vector like strategic prepositioning; and, as with lasers, the core generally-applicable approach of improved defenses to attenuate both impact and incentive coupled with explicitly readying a set of options for credible and certain imposition of proportionate costs is likely to be a winning formula.

However, this also reveals that what is most needed in the case of jamming are supplemental measures which are specifically geared toward making both the denial and response components consistently effective in the face of jamming, across a wide variety of contexts and despite rapid advances in enabling technologies (for example, increasingly-sophisticated artificial intelligence).

To that end, the authors recommend that the United States implement the following primary diplomatic and unilateral measures to aid in countering jamming:

First, the United States and its allies must able to reliably and confidently detect and attribute jamming to hostile actions by China. That requires deploying more capable space-based sensors and establishing organizational processes for better sharing and integration of private sector data. Measures to achieve this include increasing investment in comprehensive monitoring from orbit and increased integration of commercial forensics data.

Second, the United States should minimize vulnerability in three specific ways: by increasing *passive* hardening of key satellites and ground stations (e.g., controlled reception pattern antennas)[451];

451. Michael Jones, "Anti-jam technology: Demystifying the CRPA," GPS World, April 12, 2017, https://www.gpsworld.com/anti-jam-technology-demystifying-the-crpa/.

adopting *active* anti-jamming technologies (e.g., artificial intelligence-driven[452] jamming suppression); and by employing redundant platforms both in space and terrestrially.

Third, the United States should engage in close consultation with its allies and partners to build shared understandings well in advance of a crisis, and work to reach agreement regarding the legality and appropriateness of different responses under various circumstances. For example, are there circumstances under which electronic warfare could constitute a use of force? Under the conditions where it does not, when might it nevertheless rise to the level of harmful interference as understood in Art. IX of the Outer Space Treaty,[453] and what sets of lawful countermeasures does the U.S. believe are permissible under international law to enforce that provision?

Fourth and finally, the United States should clarify to potential adversaries what unilateral response options exist and might be used, including but not limited to potential in-kind responses or infliction of proportionate cross-domain costs. Furthermore, the U.S. should be consistent and swift in following through on these threats in the overwhelmingly likely event that China tests these lines.

---

452. Tracy Cozzens, "ADVA tackles GNSS jamming and spoofing with AI solution," GPS World, February 12, 2020, https://www.gpsworld.com/adva-tackles-gnss-jamming-and-spoofing-with-ai-solution/.
453. "Outer Space Treaty, Article IX."

IV. Summary of Key Recommendations

A. Specific Recommendations to Counter the Three ASAT Threats

This report recommends that the United States adopt the following specific solutions:

General: First, minimize potential costs and maximize deterrence-by-denial by accelerating adoption of measures which limit the damage that adversaries can inflict via counterspace operations. Second, maximize deterrence and minimize escalation risks by ensuring the ready availability—and clear *ex ante* articulation to potential adversaries—of an array of graduated, flexible cost-imposition measures. Third, elaborate strong international legal bases sufficient to place these response options on strong legal footing and provide clear frameworks and triggers for action; coordinate with allies *ex ante* ensure alignment of objectives and strategies and ensure that adversaries may confidently expect a smooth and unified response in a crisis; and engage with potential adversaries to build clear mutual understandings.

Targeted: First, counter strategic and operational-level threats from R-spacecraft by establishing zones around exquisite space assets and deploying bodyguard satellites to police those zones. Second, address the risks posed by both inadvertent interference and hostile RPOs by building a global STM regime through international 'rules of the road,' as well as parallel Western STM regime with conditioned market access as explained in Section III; commercial incentives for compliance; and elaborating measures for lawful retorsion in cases of violation. Third, adopt appropriate operating procedures to mitigate the threat from ground-based lasers, while moving to constrain their proliferation through international agreements and deter their use by clearly articulating available response options. Fourth, counter jamming by enhancing capabilities for attribution, making full use of advanced passive and active anti-jamming technologies, and refining options for retaliation to grey zone provocations against ourselves and allies.

B. A Synergistic Approach to Next-Generation Space Threats

Finally, and more generally, the results of this report reveal a critical lesson for policymakers and analysts as the U.S. confronts a rapidly-emerging new era in space security: the need for a **synergistic approach**. To successfully counter ASAT threats during the 2020s and beyond and maintain peace and prosperity in orbit, the U.S. will have to reject and transcend longstanding divisions between hawks allergic to diplomatic measures and doves staunchly opposed to any use of unilateral and military measures. As illustrated above, emerging space threat vectors have varied and nuanced characteristics, and dealing with them effectively requires a similarly-nuanced package of solutions. Moreover, assessing the profile of each individual counterspace threat case-by-case reveals that a package of different measures is required—some general in applicability, others narrowly-tailored to certain threats.

Unfortunately, time is not on our side. Threat identification is only the first step in a long chain of processes: deciding on a solution, developing a concept of operations, carrying out procurement, and deploying the defense. Thus far, the U.S. has been unable to complete even step two, solution selection, and the hour is growing late. This report contends that a primary driver of this glacial response is that we have been proposing solutions based on understandings which are much too general. Counterspace analysts have tended to work in silos to study China and U.S. actions and interactions and talk past one another.

With NPEC's support, this report has sought to break from that tradition. By analyzing specific threat vectors on a case-by-case basis and remaining focused on nuanced, tailored solutions, incorporating both diplomatic and unilateral measures, the U.S. has a clear path forward. The three ASAT threat vectors evaluated herein are important, and the strategic-level threat posed by unrestrained pre-positioning of R-spacecraft especially so. However, they are only the first step: if the U.S. seeks to avoid sleepwalking into strategic surprise, it must move swiftly to apply a similar frame to a number of other counterspace threats, and implement the results.

We have only a few years left; the clock is ticking.

Appendix: China's Order of Battle in Rendezvous Spacecraft for ASATs in a Taiwan Contingency by 2026

(Originally estimated for use in NPEC's China Space Wargame for a 2029 scenario in June 2021, but updated in July 2022)

Key Reasons for the Update

This appendix was originally developed to provide the order of battle (OOB) in rendezvous spacecraft (R-spacecraft) as ASATs in a 2029 scenario for a Taiwan contingency in a China Space Wargame held in June 2021. The key assumption was that China's success in docking was attained five years behind that of the U.S. In early 2021, the authors reviewed and compared the rendezvous and proximity operations (RPO) test data of China and the United States and found that the China's docking and other RPO capabilities will lag those of the United States by at most five years but likely to be shorter. However, at the time, many space experts argued that China could take far longer to attain docking. To be conservative, the authors used the five years. However, as discussed below, China succeeded such a docking in January 2022, merely two years behind the United States. As a result, whatever we estimated to be the OOB for 2029 would now be for 2026 to reflect the lag has been reduced from five to two or by three years.

Another update was to reflect China will attain faster production rate of small R-spacecraft, which will result in the OOB from 100 estimated in early 2021 to 200 estimated in July 2022.

Overview

Dual-use R-spacecraft have many unique characteristics. Accurately predicting the order of battle (OOB) for R-spacecraft by 2026 is difficult because it depends on a wide range of underlying judgements. Any specific estimate may be easily characterized as tipping the scale in favor of a specific agenda. Fortunately, exact predictions are neither necessary nor desirable for the purposes of this paper; that task may be safely left to the counterspace threat analysts in the intelligence community. Instead, all that is needed for our purposes is to establish whether Chinese production of these systems on the scale necessary to enable the counterspace operations contemplated *supra* at Section II is a realistic possibility.

Ultimately, while there is substantial uncertainty regarding the number of R-spacecraft that China *will* produce (indeed, if the premises of this paper are correct, this number is highly likely to be endogenously affected by whether or not the U.S. adopts the set of countermeasures outlined in this paper). The authors estimated that China will likely be capable of producing a quantity of R-spacecraft greater than number required to conduct the various counterspace operations contemplated *supra* at Section II, including the Taiwan scenario.

Assumptions

1. *China is highly likely to pursue R-spacecraft for ASAT use in the status quo.*

There is clear evidence that China is actively pursuing co-orbital ASAT development and testing. The most potent use of R-spacecraft by China as anti-satellite weapons (ASATs) is at the opening of a conflict because prepositioning R-spacecraft in close proximity to critical satellites could potentially deter the United States from intervening without firing a shot, or could enable sudden simultaneous strikes on the entirety of one or more strategically vital constellation(s).

2. *The OOB is likely to heavily depend on whether effective countermeasures are adopted and clearly communicated by the United States and its allies in advance.*

The quantity of R-spacecraft produced and launched by China is likely to be highly elastic: if the United States possesses an effective response and China is aware of this, then R-spacecraft are likely to be relatively unattractive for use in an ASAT role. Under such circumstances, China would likely produce few R-spacecraft despite ample production capacity. This is because China's counterspace development appears strategically motivated, other ASAT weapons are preferable for most other missions, and a much smaller number of more capable R-spacecraft are necessary and sufficient to fulfill civil and commercial RPO missions.

In contrast, if the United States and/or its allies lacks an effective response, or such countermeasures are not adopted and clearly communicated in advance, then there is a moderate to high probability that China's OOB for R-spacecraft will be large. This is a function of both capability and will. First, as outlined below, the R-spacecraft on which China is focusing are cheap and China has the capacity to produce many in short order. Second, China is well aware of the unique strategic opportunities such prepositioning would afford and has demonstrated interest in such operations.

3. *China is likely to have an operational capability to enable multiple R-spacecraft to dock and disable our satellites by 2026.*

In February 2020, Northrop Grumman's commercial R-spacecraft, Mission Extension Vehicle-1 (MEV-1) for the first time in history autonomously docked with Intelsat-901. It is important to note that MEV docked with a satellite that was not pre-designed[454] with docking in mind. "MEV-1 autonomously reached out with its docking probe, that entered the nozzle of the liquid apogee engine on Intelsat 901. The docking probe was inserted into the engine nozzle and through the "throat" — narrowest part of the engine nozzle. Once the top of the probe was beyond the throat, it deployed a series of "fingers" to expand its size — like a wall anchor once inserted into a wall."[455] This means that if China were to attain the same or alternative capability, it could "dock" with a U.S. satellite take it to a useless orbit. Once the United States complied with China's demand (e.g., no intervention), China could bring the captured satellite back to its original orbit unharmed. Alternatively, once "docked," the Chinese R-spacecraft could use its robotic arm to bend and/or disconnect antennas and solar panels, etc. to disable our satellites without creating much space debris.

As discussed in Section II, China's Shijian-21 docked with a defunct BeiDou-2 G2 navigation satellite in January 2022,[456] By 2026, it should attain a capability of operating multiple R-spacecraft, as these operations are independent of each other in terms of controlling them in orbit and/or on the ground.

---

454. Caleb Henry, "Northrop Grumman's MEV-1 servicer docks with Intelsat satellite," SpaceNews, February 26, 2020, https://spacenews.com/northrop-grummans-mev-1-servicer-docks-with-intelsat-satellite/ .

455. Chris Gebhardt, "Northrop Grumman makes history, Mission Extension Vehicle docks to target satellite, NASASPACEFLIGHT.COM, February 26, 2020, https://www.nasaspaceflight.com/2020/02/northrop-grumman-history-mission-extension-vehicle-docks-satellite/

456. Evan Gough, "A Chinese space tug just grappled a dead satellite," Phys.org, February 2, 2022, https://phys.org/news/2022-02-chinese-space-grappled-dead-satellite.html

4. Satellites suitable for co-orbital ASAT use can be substantially smaller and cheaper than the MEVs.

Whereas MEV-1 has a mass of 2,326 kg[457] and MEV-2, 2,875 kg[458] for commercial applications, China appears to optimize its R-spacecraft for ASAT use. Both Sheyann and Aolong-1,[459] used as Chinese experimental spacecraft, as well as similar Russian spacecraft (e.g., Kosmos 2499 and Luch),[460] are small satellites, not exceeding 200 kg[461] each. As for the SJ-12 and SJ-06F used in some RPO tests, there is no data on their masses. However, they belong to the Shi Jian class and six of them have open data, showing their masses ranging from 221 kg to 480 kg.[462] Therefore, it is likely that both SJ-12 and SJ-06F would be in this mass range.

For context, it is reasonable to assume that the cost of a small satellite (i.e., less than 200 kg) is on the order of $10 million or less,[463] depending on the payload. This is far less expensive than their potential targets—traditional U.S. satellites such as early warning satellite ($1.7 billion)[464], satellite for communication under nuclear-disrupted environment ($1.7 billion),[465] or GPS satellite ($577 million)[466] cost approximately 50-150x more. To be

---

457. "MEV 1,2,3," Gunter's Space Page, https://space.skyrocket.de/doc_sdat/mev-1.htm .
458. Ibid.
459. Bhavya Lal, etc., *Global Trends in Small Satellites*, IDA, July 2017, E-13, https://www.ida.org/-/media/feature/publications/g/gl/global-trends-in-small-satellites/p-8638.ashx#page=249 .
460. Lal, "Global Trends in Small Satellites," E-20.
461. Lal, "Global Trends in Small Satellites," iii.
462. Ibid.
463. Chow, "Space Traffic Management in the New Space Age," 7.
464. Chow, "Nuclear vulnerability."
465. Ibid.
466. Dan Elliott, "Next-generation of GPS satellites are headed to space," Phys.org, December 17, 2018, https://phys.org/news/2018-12-next-generation-gps-satellites-space.html#:~:text=The%20price%20of%20the%20first,adjusted%20for%20infl

conservative, the authors rounded them to 10-100x, when this range was used in Section III.

5. *China is highly likely to have the technological prowess to both produce and launch R-spacecraft in numbers sufficient to threaten critical assets in GEO, HEO, and MEO (i.e., GPS).*

There are indications that China is showing significant progress in producing a large number of small satellites, including R-spacecraft. In January 2021, Global Times[467] reported that China's first intelligent mass production line is set to produce 240 small satellites per year. It took China Aerospace Science and Industry Corporation (CASIC) only 429 days to complete the line. This line increased production efficiency by over 40 percent during 2021, and began trial operations.

In April 2022, production trials were underway at the China Academy of Space Technology (CAST), which will be capable of producing more than 200 satellites per year. As discussed in this appendix and Section II, China has succeeded docking in January 2022. Then, companies are likely to enable mass-production of payloads and satellites, and it is conservative to assume that, by 2026, China can have produced 200 R-spacecraft if it decides to do so.

This should surprise no one: China's investments in developing sophisticated space capabilities have been consistent, large, and fruitful. In 2014, Beijing opened its space industry to private companies, emphasizing small satellites. In September 2019, the

---

ation%2C%20Chaplain%20said.&text=The%20Air%20Force%20said%20in,the%20cost%20at%20%2412%20billion .

467. "China's first intelligent mass production line for satellites able to produce 240 small satellites each year," Global Times, January 20, 2021, https://www.globaltimes.cn/page/202101/1213345.shtml#:~:text=As%20China%20ramps%20up%20efforts,240%20small%20satellites%20each%20year .

Science & Technology Policy Institute at IDA released a major report on *Evaluation of China's Commercial Space Sector* (Irina Liu, et al). Their analysis included a high-level overview of the 78[468] commercial space companies, 28 of which are in the sector of satellite and component manufacturing and 5 more in satellite navigation. On the counterspace side, China has been systematically pursuing offensive space capabilities for potential use against the United States since the First Gulf War during August 2, 1990-February 28, 1991 over three decades ago, and the fruits of their labor have been obvious in recent years.

We should expect that if China assesses R-spacecraft to be important for their asymmetric counterspace and military strategy, they will dedicate the resources necessary to ensure that the private companies can produce the number of R-spacecraft they need. According to the Space Foundation,[469] the Chinese military and government have an $8 billion annual budget. Yet, 200 small R-spacecraft at $10 million each would amount to $2 billion; a highly worthy investment considering it would allow China to disable over a 100 U.S. critical satellites at GEO, HEO, and MEO at the opening of a war.

As demonstrated by the China's use of small satellites for R-spacecraft and for promoting commercial manufacturing and applications in small satellites, it appears that China is undertaking all preliminary work necessary to enable mass-production of payloads and satellites for R-spacecraft by 2026.

---

468. Irina Liu etc., *Evaluation of China's Commercial Space Sector*, IDA, September 2019, iv, https://www.ida.org/-/media/feature/publications/e/ev/evaluation-of-chinas-commercial-space-sector/d-10873.ashx#page=6 .
469. Blake Schmidt, A Millennial's Tiny Satellites Are Helping China Advance in the Space Race, Bloomberg, December 13, 2018, https://www.bloomberg.com/news/features/2018-12-13/a-millennial-s-tiny-satellites-are-helping-china-advance-in-the-space-race .

Table A.1: # of Chinese Rendezvous Spacecraft for Peaceful Use

Year of Conflict	# of Rendezvous Spacecraft
2026	8

Basis for Our Projection:

1. As outlined above, we conservatively assume that China's RPO operational capability lags that of the United States by 2 years.
2. Northern Sky Research (NSR)[470] recently forecast that demand for GEO life extension missions will grow to 75 satellites by 2030, representing a $3.2 billion market. We assume that life extension missions will include refueling, repairing, upgrading and relocating services. We assume that the number of space debris removal missions is also 75 by 2030. This produces an estimate of 150 missions annually by 2030 for R-spacecraft worldwide to perform for satellite servicing and debris removal. We also assume that by 2030 that these 150 missions will be provided by 75 R-spacecraft worldwide. Finally, we assume that the global market for satellite servicing and debris removal splits 4 ways: United States, European Union (EU), China, and the rest of the world each get 25% of the 150-mission demand and provide 25% of the 75 R-spacecraft required. This would mean that

470. Jason Rainbow, "Northrop's MEV-2 servicer closing in on Intelsat-10-02 docking attempt," SpaceNews, March 10, 2021, https://spacenews.com/mev-2-servicer-closing-in-on-intelsat-10-02-docking-attempt/ .

China would have or require on the order of 19 R-spacecraft by 2030.

3. We assume that China will add equal number of R-spacecraft annually starting 2023 so as to reach 19 R-spacecraft by 2030. This produces the following estimates: 8 spacecraft by 2026. Moreover, China would likely claim R-spacecraft intended for ASAT purposes as for peaceful uses until they actually attack our satellites.

Table A.2: # of Chinese Rendezvous Spacecraft Intended for ASAT Purposes

Year of Conflict	Low # of ASAT Rendezvous Spacecraft	High # of ASAT Rendezvous Spacecraft
2026	0	200

Basis for Our Projection:

1. The lower-bound estimate assumes that the United States and its allies adopt clearly effective countermeasures in the near future, thereby rendering prepositioning of R-spacecraft ineffective and undesirable. As mentioned above, under such a scenario it is plausible that China would lose any desire to employ R-spacecraft for counterspace operations in the timeframe discussed.
2. The upper-end estimate, in contrast, assumes that U.S. responses to the rendezvous threat follow the current trend of pursuing some necessary measures in all three arenas (i.e., international, multilateral and unilateral responses), but in a manner that is insufficient and/or untimely. Based on the discussion in this appendix about that China has already

attained the docking and other RPO capabilities and what it has achieved in manufacturing capability of small satellites, it is conservative to assume that China would be capable of producing 200 R-spacecraft for ASAT purposes, very likely more.

The Anti-Satellite (ASAT) Consequences of a Nuclear Explosion in Space

Gregory S. Jones[471]

Introduction

On July 10, 1962, the first active communications satellite, Telstar, was launched into low Earth orbit (LEO). On the day before Telstar's launch, the U.S. conducted a high-altitude nuclear explosion (HANE). This was the Starfish Prime shot, of the Dominic test series. The device had a yield of 1.4 megatons and was exploded at an altitude of 400 kilometers. It was carried on a Thor medium-range ballistic missile, which had a surface-to-surface range of 2,400 km.

Within days of the Starfish Prime explosion, there were multiple satellite failures. One of these was Ariel-1, which was the UK's first satellite. It stopped functioning on July 13, 1962. Telstar lasted longer but stopped functioning in November 1962. Altogether, eight satellites were damaged by the Starfish Prime test, fully one-quarter of all satellites in orbit at the time. The satellites were not damaged by the direct effects of the nuclear explosion (X-rays and electromagnetic pulse [EMP]) but rather by high energy electrons (beta

471. This paper was produced for the Nonproliferation Policy Education Center. Though the author is also a part-time adjunct staff member at the RAND Corporation, this paper is not related to any RAND project and therefore RAND should not be mentioned in relation to this paper. I can be reached at GregJones@proliferationmatters.com

particles) released by the nuclear device's fission products and trapped by the Earth's magnetic field.

Today there are nearly 5,500 active satellites in orbit and the U.S. and many other countries are far more dependent on such space assets than they were in 1962. These satellites perform important communications and observation functions. In addition, there is a continuous human presence in space as part of space exploration efforts. A nuclear explosion in space could disable many of these satellites leading to large direct and indirect economic costs. Such an explosion could be the result of a variety of actions, such as a deliberate anti-satellite attack, a high altitude "warning shot" in an ongoing conflict or a salvage-fused warhead intercepted by an ABM.

This paper will examine the mechanism whereby nuclear explosions disable satellites, the types of orbits in which the satellites would be disabled and the type of orbits in which they would not, the locations above the Earth where a nuclear explosion would have the most impact, the direct and indirect economic impact of the destruction of satellites, the effectiveness of standard ASAT countermeasures and the overall uncertainties involved.

The Van Allen Belts and Satellite Orbits

The Earth's magnetic field traps protons and electrons that stream outward from the sun. These trapped particles form two distinct radiation belts, which are known as the Van Allen belts. These protons and electrons can damage the electronics on satellites and as such they form a serious constraint on the location of satellite orbits. The inner belt generally extends between 2,500 km to 6,500 km above the Earth. The outer belt is larger and stronger. It extends between about 12,000 km and 35,000 km above the Earth.

Fluctuations in the solar winds can significantly affect the intensity and upper bound of the outer belt. The Van Allen Belts are shown in Figure 1.[472]

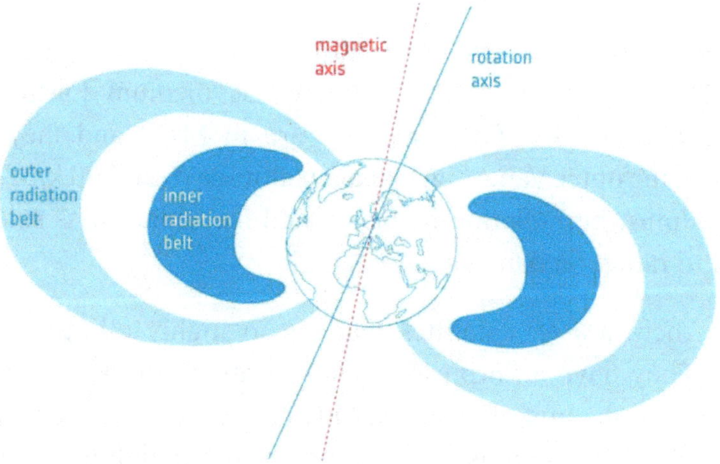

Figure 1: Van Allen Radiation Belts

Most active satellites (4,700 out of 5,465 as of April 30, 2022), are in low Earth orbit (LEO). They orbit no higher than 2,000 kilometers.[473] This is a benign region below the innermost Van Allen belt and is protected, since most space radiation is trapped in the Van Allen belts.

Most of the remaining satellites (565) are in geostationary Earth orbit (GEO). These satellites orbit at an altitude of 35,800 km. They

472. The European Space Agency https://www.esa.int/ESA_Multimedia/Images/2016/06/Illustration_showing_Van_Allen_radiation_belts
473. Union of Concerned Scientists Satellite Database https://www.ucsusa.org/resources/satellite-database

have an orbital period which exactly matches the Earth's rotational period (23 hours and 56 minutes). These satellites always remain over the same part of the Earth. Their orbits are near the outer edge of the outer Van Allen belt and these satellites require additional radiation hardening compared with satellites in LEO.

There are also some satellites (140) that orbit between LEO and GEO. These satellite orbits are known as medium Earth orbits (MEO). The 24 U.S. GPS satellites are in MEO and they orbit directly in the outer Van Allen belt at an altitude of 20,200 km. These satellites must be radiation-hardened and are replaced as satellites are lost to radiation damage.

Finally, there are 60 satellites that are in highly elliptical orbits (HEO). They have an orbit low point of about 900 km above the Earth and a high point of about 39,000 km. Due to orbital mechanics, these satellites spend most of their time near the high point of their orbits and move quickly through their low point. These satellites can function similarly to satellites in GEO, providing communications to or observations of a fixed area of the Earth. It takes a constellation of three such satellites to provide continuous coverage of a particular area. These satellites have the advantages of being able to cover polar areas, where the coverage provided by GEO satellites is poor and they require significantly less propulsive energy to place into orbit compared to a satellite in GEO.

HANEs and Enhanced Radiation Belts

HANEs create their own belts of enhanced radiation. The fission products produced by a nuclear explosion decay by the emission of a beta particle which are simply electrons, many of which have high energy. These electrons are captured in the Earth's magnetic field where they persist for some time.

The Earth's magnetic field can be considered a dipole. The magnetic field lines exit the surface of the Earth somewhere in the Southern Hemisphere and head towards the magnetic equator. They increase in altitude until they reach their greatest height over the magnetic equator and then reenter the Earth somewhere in the Northern Hemisphere. These magnetic field lines are termed "L-shells." They are characterized by the distance from the center of the Earth where they cross the magnetic equator measured in terms of the Earth's radii. For example, an L-shell equal to 2 passes over the magnetic equator two Earth radii above the Earth's center, which is one Earth radii (6,371 km) above the Earth's surface. L-shells are illustrated in Figure 2.[474] Note that the higher numbered L-shells exit and enter the Earth closer to the Earth's poles than do the lower numbered L-shells.

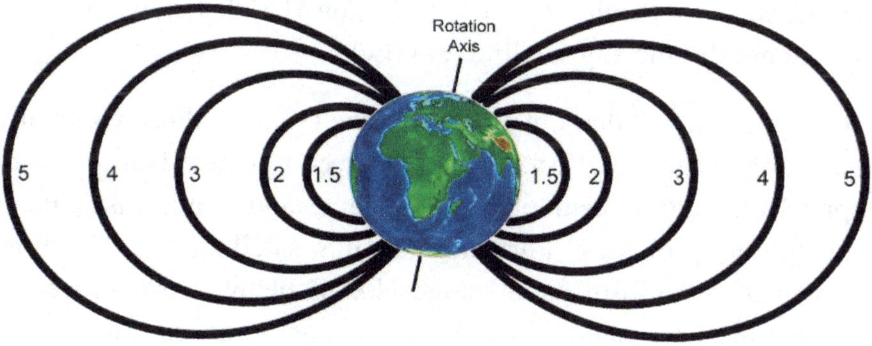

Figure 2: L-shells

The electrons from a HANE fill tubes of space confined by the L-shells. The altitude and geomagnetic latitude of the HANE determines which L-shells are filled with radiation. My main focus

474. Wikipedia https://en.wikipedia.org/wiki/L-shell

will be on a HANE below 2,000 km i.e., no higher than LEO. The L-shells filled from such a HANE depend on its geomagnetic latitude and could affect satellites in either GEO, MEO or LEO.

A HANE below 2,000 km near the Earth's poles (55 to 70 degrees latitude north or south) would fill L-shells of around 6.6 and would cause enhanced radiation at GEO. A HANE below 2,000 km somewhat closer to the equator (45 to 65 degrees north or south) would fill L-shells of around 4 and would affect the GPS satellites. A HANE below 2,000 km above the equatorial region (between about 35 degrees north and 35 degrees south) would fill L-shells of 1.3 or below and would affect satellites in LEO.

A HANE below 2,000 km directed at satellites in GEO would have little effect as the enhanced radiation belt would be distributed into a large volume and the radiation belt would disperse in about a month.[475] Further, the satellites in GEO are somewhat radiation-hardened. As a result, even a 10 megaton HANE below 2,000 km would not disable any satellites at GEO.

A HANE below 2,000 km directed at the GPS satellites would also have only a limited effect.[476] The enhanced radiation belt would be spread out into a relatively large volume and would dissipate in roughly four months. Further, the GPS satellites are radiation-hardened. A multi-megaton range HANE below 2,000 km could shorten the GPS satellite life from ten years to seven. Since some GPS satellites are near the end of their lives, after some time, a few could fail. Initially, spare satellites would be able to fill the gaps but over time some gaps in coverage could occur if steps are not taken

---

475. Edward E. Conrad et. al., "Collateral Damage to Satellites from an EMP Attack," Defense Threat Reduction Agency, DTRA-IR-10-22, August 2010, p. 105. https://apps.dtic.mil/sti/citations/ADA531197
476. *Ibid.*

to deal with this problem. However, it would be at least one year after the HANE before this became an issue.

A HANE below 2,000 km with a yield of just 10 to 30 kilotons, directed at satellites at LEO, could have a devastating effect.[477] The enhanced radiation belt would be confined to a small volume and the enhanced radiation could last for years. Satellites in LEO are not designed for high radiation and the enhanced radiation would be 10,000 times higher than normal. In a few months, most satellites in LEO would fail. Many might fail in days or weeks.

Figure 3 shows the difference in radiation intensity between a HANE at a high geomagnetic latitude and one at a low geomagnetic latitude.[478] The radiation field from a HANE at a high geomagnetic latitude reaches high above the Earth as it fills high numbered L-Shells and can reach satellites in GEO or MEO. However, the radiation field is dispersed into a large volume and is relatively weak. The radiation field from a HANE at a low geomagnetic latitude fills low numbered L-Shells much closer to Earth and would reach satellites in LEO. The radiation field would be contained in a small volume and would be relatively high. The high radiation field, plus the fact that satellites in LEO are not hardened against radiation effects, means that these satellites would be quite vulnerable to a HANE at a low geomagnetic latitude.

---

477. *Ibid.*, pp. 103-104.

478. Johan Idestrom, "Nuclear hardening to protect satellites against high-altitude-nuclear-explosions (HANE)," OHB System AG, November 8, 2018, p. 15. https://documents.pub/document/nuclear-hardening-to-protect-satellites-against-high-altitude-2018-11-08-leo.html?page=1

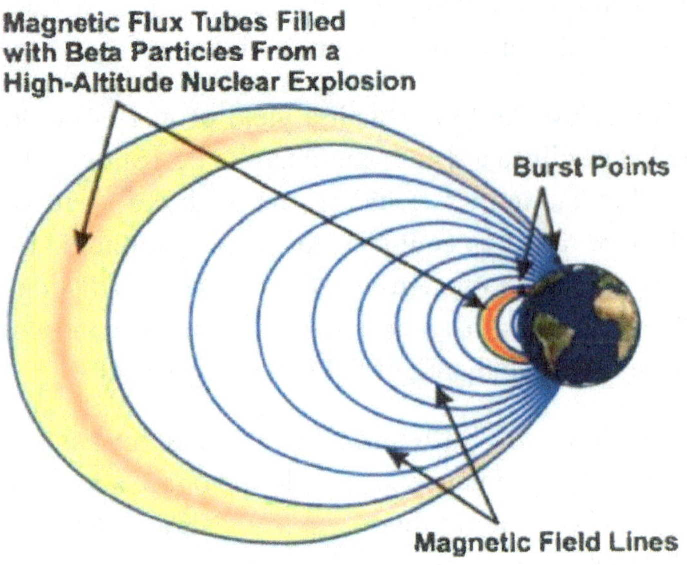

Figure 3: Difference in the Intensity and Altitude of the Radiation Field from a HANE at a High Geomagnetic Latitude and One at a Low Geomagnetic Latitude

A HANE below 2,000 km with a yield of just 10 to 30 kilotons, directed at satellites at LEO would not seriously affect HEO satellites. Such satellites do not spend much time below 2,000 km, since, as was stated above, they instead spend most of the time at the high point of their orbits which is near GEO. However, a high yield HANE of 5 megatons, exploded below 2,000 km, would increase the radiation belt in LEO sufficiently so as to disable HEO satellites in about one month.[479]

A HANE below 2,000 km would also generate direct effects (x-rays and EMP) that could promptly destroy satellites. For weapons with

479. Edward E. Conrad et. al., "Collateral Damage to Satellites from an EMP Attack," Defense Threat Reduction Agency, DTRA-IR-10-22, August 2010, p. 109. https://apps.dtic.mil/sti/citations/ADA531197

yields in the 1 megaton to 10 megaton range, these effects could extend out for several thousand kilometers. However, due to the shielding of the Earth, at most only 5% to 10% of a LEO satellite constellation would be destroyed. Also, a HANE below 500 km could generate substantial EMP damage to electrical equipment on the Earth. The damage would extend to an 800 km to 1,600 km radius from the point directly below the explosion.

For a HANE that is above 2,000 km, the enhanced radiation belts will only affect satellites in MEO or GEO. The impact of such enhanced radiation belts on satellites in these orbits will be no more severe than from that of a HANE that is below 2,000 km. As was discussed above the resulting radiation belt from such a HANE would not damage satellites at GEO and do only minimal damage to GPS satellites. However, there are two further issues related to HANEs above 2,000 km.

First, it is difficult for a missile carrying a nuclear warhead reach altitudes much above 2,000 km. The 2,400 km Thor medium-range ballistic missile mentioned in the introduction would only reach a height of around 1,200 km if fired straight up.[480] Even a 11,000 km intercontinental-range ballistic missile carrying its nominal payload would only reach an altitude of about 5,000 km if fired straight up. This altitude is still well below the GPS satellites, let alone the ones at GEO. However, if a country had a MIRVed ballistic missile, it could off-load all but one of the warheads. The single remaining warhead would be able to reach escape velocity and therefore reach any altitude desired.

Second, the Starfish Prime shot discussed in the introduction was the highest altitude nuclear explosion to ever occur. Though the direct effects of a high-altitude nuclear explosion (x-rays and EMP)

---

480. These are the author's calculations. See: Roger R. Bate, Donald D. Mueller and Jerry E. White, *Fundamentals of Astrodynamics*, Dover Publications Inc., New York, 1971, Chapter 6.

at MEO or GEO would probably be able damage satellites, there are substantial uncertainties in the effectiveness of a nuclear explosion at these high altitudes. Despite the uncertainties it is likely that a 1 megaton to 10 megaton nuclear weapon could destroy satellites out to one or two thousand kilometers away from the blast point. However, satellites are far apart at MEO or GEO and a HANE at these altitudes would probably only disable a handful of satellites. Further, the countries of greatest concern such as North Korea, India, Pakistan or Iran are unlikely to have MIRVed ICBMs nuclear weapons with yields in the megaton range for many years to come.

Impact of the Loss of Satellites in LEO

The satellites in LEO perform two main functions. First, the majority of the satellites perform some sort of communications function. This can take the form of telephone service but most of these satellites actually perform a data transfer function. The latest version of this data transfer is the internet service that is being provided by the Starlink satellites. Second, LEO satellites provide Earth observation services, including weather satellites. In addition to these two functions, all current human space flight, including the International Space Station, occurs in low Earth orbit.

A detailed economic analysis of the impact of the loss of all satellites in LEO, which currently number 4,700 and could soon be in the tens of thousands, is beyond the scope of this paper. Instead, I will look at a few sample satellite constellations and make a rough extrapolation to the entire number of LEO satellites.

More than half of the satellites currently in LEO are part of Elon Musk's Starlink program. Starlink uses a large number of small (500 lb.) satellites in LEO to provide internet service. Currently there are around 3,000 Starlink satellites in orbit and service has begun with hundreds of thousands of subscribers. The Starlink program intends to ultimately have 12,000 satellites in orbit for its first phase. The cost of building each satellite plus the cost of placing it in orbit are reported to be under $1 million. Therefore, the cost of the full 12,000 satellite constellation would be about $10 billion.

The economic value of the Starlink program could be significantly higher. Musk has predicted that by 2025, Starlink would generate $30 billion a year in revenue.[481] This would make providing internet service a far larger source of income for SpaceX than its space launch services. This claim has met with some skepticism but Starlink could easily generate at least $10 billion a year.

Nor is Starlink the only satellite internet program. Amazon plans to deploy over 3,000 satellites as part of its Project Kuiper. It has already contracted for $10 billion worth of launch services with the first launches to occur in the later part of 2022.

An example of an Earth observation satellite program is the European Space Agency's Copernicus program, which consists of about a half a dozen Sentinel satellites.[482] These satellites carry a variety of instruments such as synthetic aperture radar, and radiometers, as well as multispectral visible and infrared optical sensors. The purpose of these satellites is to monitor environmental conditions on the Earth including the oceans. The satellites are significantly heavier (3,000 lb. to 5,000 lb.) than the Starlink satellites, which means that even this small constellation has involved significant costs. It has been reported that over $8 billion

---

481. Rich Smith, "Elon Musk Predicted Starlink Would Generate $ 30 Billion in Revenue. Don't Believe It. February 14, 2021.
https://www.nasdaq.com/articles/elon-musk-predicted-starlink-would-generate-%2430-billion-in-revenue.-dont-believe-it.-2021

482. The number varies over time as some satellites are retired and others launched to replace them.

has been invested in the program thus far.[483] The data is provided for free so there is no direct economic income from Copernicus.

By far the most valuable object in LEO is the International Space Station (ISS). It is reported to have cost $150 billion and may be the most expensive single item ever constructed. A HANE would require the immediate evacuation of the ISS. Over time, the electronics onboard would be destroyed, which could make the ISS a total loss. A further problem is that the ISS is in a low enough orbit that atmospheric drag causes the orbit to drop. Under normal circumstances, burns are performed to raise the orbit but this may not be possible after a HANE and the ISS could burn up in the Earth's atmosphere.

In sum, the satellites in LEO represent substantial value. Just the internet services of Starlink and Amazon could be worth many tens of billions of dollars. Plus, there are many other communication satellites in LEO. In addition to the six Sentinel Earth observation satellites, there are roughly another one hundred. Their total value will certainly also be in the many tens of billions of dollars. Adding in the $ 150 billion for the ISS, the total value would roughly be between $200 and $300 billion.

Nor is all of the value of the satellites in LEO simply economic. Many of the satellites perform important military functions such as both communications and Earth observation functions (i.e., "spy" satellites). The degree to which these military satellites are radiation-hardened is unknown. In some cases, the military relies on civilian satellites for these services and most all such satellites would cease to function soon after a HANE that affected LEO.

---

483. Alessandra Tessa, "The socio-economic value of satellite earth observations: huge, yet to be measured," *Journal of Economic Policy Reform*, Vol. 23, No. 1, 2020. p. 36.
https://www.tandfonline.com/doi/full/10.1080/17487870.2019.1601565

One example of a possible military impact of a HANE below 2,000 km and at a low geomagnetic latitude, relates to the U.S. ballistic missile early warning system, Space-Based Infrared System (SBIRS). This system consists of six satellites, four in GEO and two in HEO. As was discussed above, even a 10 megaton HANE directed at GEO would have little effect on the four satellites in GEO. However, since the two satellites in HEO spend some time below 2,000 km, a HANE with a 5-megaton yield directed at LEO might disable these two satellites in one month.

Countermeasures Against Conventional ASATs Will Not Protect Against a HANE

Since 1985, the U.S., Russia, China and India have all used direct assent weapons to destroy target satellites in orbit. Older satellite systems are especially vulnerable to such an attack since due to the satellites being large and expensive, they involved only a small number of satellites. The direct assent interceptors would cost far less than the satellites destroyed.

A proposed countermeasure against this sort of threat would be to deploy a satellite system involving a large number of small satellites which would operate as a distributed network instead of a small number of large satellites. In this way, if a few direct assent interceptors were to destroy several of these smaller satellites, the system would continue to function. Unfortunately, this countermeasure would be ineffective against a HANE below 2,000 km at a low geomagnetic latitude affecting satellites in LEO, since all of the satellites in the distributed network would be destroyed. Indeed, it may be more difficult to harden these smaller satellites against the effects of radiation than to harden a handful of large satellites.

Another proposed countermeasure against conventional ASAT attacks would be to quickly replenish destroyed satellites. However, this countermeasure will be ineffective against a HANE. Since the high radiation levels from a HANE below 2,000 km at a low geomagnetic latitude will last for several years, the rapid launching of replenishment satellites will only result in those satellites failing as well. LEO may be unusable for six months to two years.

Figure 4, which appeared in a Defense Threat Reduction Agency report illustrates the problem.[484] This study looked at three different satellite systems, Globalstar, Iridium and NOAA weather satellites and assumed that each satellite system was radiation-hardened to twice the level of radiation that it would receive in normal operation. In order for a replacement satellite to survive for its full-service life, one would need to wait nearly six months to launch new Globalstar satellites even though such satellites were assumed to be the most radiation-hardened of the three systems. For NOAA satellites, the wait would be more than 24 months.

484. "High Altitude Nuclear Detonations (HAND) Against Low Earth Orbit Satellites ("HALEOS"), Defense Threat Reduction Agency, April 2001, p. 16. https://spp.fas.org/military/program/asat/haleos.pdf

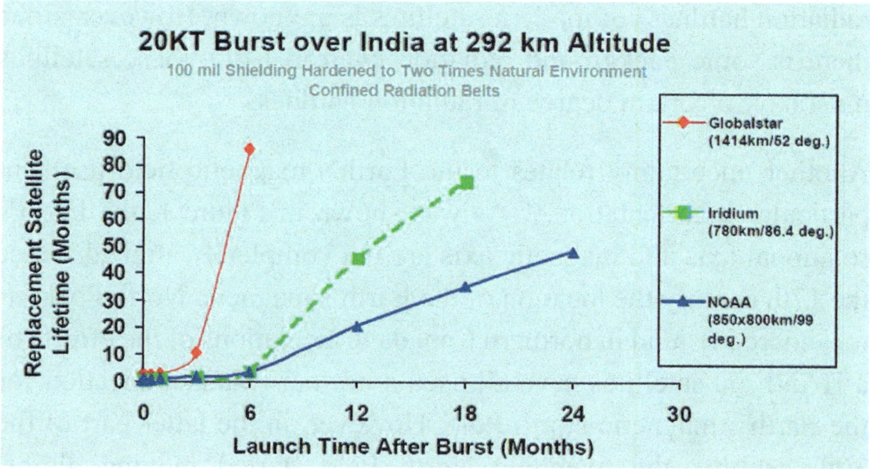

Figure 4: Time Required Before Replacement Satellites Can be Launched After a HANE

Uncertainties

Note there are substantial uncertainties regarding the impact of a HANE below 2,000 km at a low geomagnetic latitude on satellites. There have been no HANEs since 1962. Even for the HANEs that occurred in the late 1950s and early 1960s, the altitude of detonation was no higher than 400 km. The results of these tests can be only extrapolated up to about 600 km. As a result, there are substantial uncertainties about the effect of a HANE detonated between 600 km and 2,000 km.[485] Due to the proximity of the atmosphere, it is not clear how effective a HANE exploded below 90 km will be.[486]

In addition, the design of satellites and the electronics on them have changed substantially since the 1962 Starfish Prime HANE. The

---

485. Edward E. Conrad et. al., "Collateral Damage to Satellites from an EMP Attack," Defense Threat Reduction Agency, DTRA-IR-10-22, August 2010, p. 134. https://apps.dtic.mil/sti/citations/ADA531197
486. *Ibid.*, p.136.

radiation hardness of modern satellites is unknown. However, since there is some background radiation even in LEO, these satellites must have a certain degree of radiation hardness.

Another uncertainty relates to the Earth's magnetic field itself, in particular its orientation.[487] As was shown in Figure 1, the Earth's rotational axis and magnetic axis are not completely aligned. Since the 17th century the location of the Earth's magnetic North Pole has meandered around in northern Canada. Calculations of the effects of a HANE on satellites have all used a northern Canada location for the Earth's magnetic North Pole. However, in the latter part of the 20th century, the magnetic North Pole started moving almost straight north and this movement accelerated in the 21st century. Currently, the magnetic North Pole is within three and one half degrees of the Earth's rotational North Pole and has moved into the Eastern hemisphere. This movement has changed the geomagnetic latitude of locations on the Earth. For example, the geomagnetic latitude of the Korean peninsula has significantly increased. A HANE over Korea 50 years ago might have affected satellites in LEO but it is not clear that this would be the case today.

Another significant uncertainty concerns the disabled satellites themselves. Since they will be no longer under control, how likely are they to begin colliding with each other and generating space debris? The problem is only likely to grow once tens of thousands

487. Philip W. Livermore, Christopher C. Finlay, and Matthew Bayliff, "Recent north magnetic pole acceleration towards Siberia caused by flux lobe elongation," *Nature Geoscience*, Vol. 13, May 2020. https://dyz6l42c0kkca.cloudfront.net/articles/downloaded/nature/87af09c17da0ca60e98b0c7b301b53683d1688ae7aa02e797ea2276c95e5e653.pdf?response-content-disposition=attachment%3Bfilename%3D%22Livermore_et_al-2020-Nature_Geoscience.pdf%22&Expires=1665619200&Signature=e8oyhFt--rvH9gLTr9YpPNU6ixCHkI8eQz0xpRtcso0A68S1VPuA6G-nNL0yfHg1bCeQDCAa0XmLDz8F8RVhD-wdyXP0INq2aRTwZh4emHWwj6NDcl-UM9MioUOBLXY8IOovlOO2ESksyizIh0mtrqjwIU~KtonjmMmBIt~1CD8_&Key-Pair-Id=APKAI2AQJBOTGLBL6N3A

of small satellites are placed in LEO. In the worst case, the use of LEO might be permanently impaired by large quantities of space debris.

Conclusions

A HANE below 2,000 km at a low geomagnetic latitude with a yield of 10 to 30 kilotons could cause most of the satellites in LEO to fail in days to months after the explosion. LEO could be unusable for six months to two years after the explosion. A similar HANE with a yield of 5 megatons could destroy most satellites in HEO in about one month.

The economic value of the lost satellites and their services is roughly between $200 billion and $300 billion. The military satellites in LEO perform both communications and Earth observation functions. The degree to which these military satellites are radiation-hardened is unknown. In some cases, the military relies on civilian satellites for these services and most all such satellites would cease to function soon after a HANE that affected LEO. For high yield HANE, military satellites in HEO could be affected as well.

The standard countermeasures against conventional ASATs attacks such as the creation of distributed networks of small satellites or the quick replenishment of destroyed satellites will be ineffective against a HANE below 2,000 km affecting LEO. All of the satellites in a distributed LEO network will be destroyed and small satellites will probably be more vulnerable to the enhanced radiation of a HANE than would be a smaller network of larger satellites. Since the high radiation levels from the HANE below 2,000 km at a low geomagnetic latitude will last for several years, the rapid launching of replenishment satellites will only result in those satellites failing as well. LEO may be unusable for six months to two years.

The HANE would not even need to occur with the intent to deliberately to destroy satellites. Rather, it could be collateral damage from a nuclear conflict, where the HANE could be a warning shot or a salvaged fused ABM intercept.

There are substantial uncertainties regarding the consequences of a HANE exploded below 2,000 km at a low geomagnetic latitude on satellites. There have been no HANEs since 1962. The design of satellites has changed substantially since that time and the hardness of modern satellites is unknown. Also, since the highest HANE was only 400 km above the Earth, the effect of HANEs at an altitude between 600 km to 2,000 km are uncertain. Due to the proximity of the atmosphere, it is not clear how effective a HANE below 90 km will be. The rapid movement of the Earth's magnetic pole in the last fifty years has changed the geomagnetic latitude of places on the Earth and the results of some prior analysis regarding a HANE's effect on satellites in LEO may need revision. Another significant uncertainty concerns the disabled satellites themselves. Since they will be no longer under control, how likely are they to begin colliding with each other and generating space debris?

Given the growing nuclear arsenals in countries such as India, Pakistan, North Korea and in the future possibly Iran, serious study is required to determine how to best handle the threat of a HANE destroying most of the satellites in LEO.

Current Space Law: How Much Protection Might It Provide?

Laura Montgomery

The Outer Space Treaty of 1967[488] (the "Outer Space Treaty" or the "Treaty") was entered into during the height of the Cold War and during the space race to the Moon between the United States and the Soviet Union. The Treaty establishes the basis for a legal regime governing the activities of nation-states, referred to as "States Parties" throughout the Treaty, in outer space.[489] Under Article XIII, the Treaty applies "to the activities of States Parties to the Treaty in the exploration and use of outer space." As the U.S. Secretary of State, Dean Rusk, explained in his statement to the Senate, "[t]he treaty is not complete in all possible details. It does not deal with all problems that may develop. But it is responsible to those problems that can be described and forecast today."[490] This assessment focuses on whether and how the Treaty prohibits military activities in outer space, with an eye toward determining which are clear and which are ambiguous.

In summary, it is clear that States party to the Treaty may not place nuclear weapons or other weapons of mass destruction in a full orbit around the Earth. They may deploy intercontinental ballistic

488. Treaty on Principles Governing the Activities of States in the Exploration and Use of Outer Space, including the Moon and Other Celestial Bodies (the "Outer Space Treaty"), Jan. 27, 1967, 610 U.N.T.S. 205 (adopted by the General Assembly in its resolution 2222 (XXI) and entered into force on 10 October 1967.

489. Treaty on Outer Space: Hearing before the Comm. on Foreign Relations of the U.S. Senate, 4, 90th Cong. (1967).

490. Treaty on Outer Space: Hearing before the Comm. on Foreign Relations of the U.S. Senate, 4, 90th Cong. (1967).

missiles ("ICBMs") armed with nuclear warheads. Although the weight of opinion of two major spacefaring nations appears to be that fractional orbital bombardment systems armed with nuclear warheads comply with the Treaty, one may argue that that conclusion is not certain. States party to the Treaty may place weapons that are neither nuclear nor WMD in orbit, on celestial bodies, including the Moon, or otherwise in outer space. Similarly, there is no Treaty prohibition on military reconnaissance, communications, or navigation satellites. The Treaty requirement to use celestial bodies exclusively for peaceful purposes allows for self-defense. Space between Earth and the Moon does not fall under this requirement, so it need not be used exclusively for peaceful purposes. Parties to the Treaty undertake not to establish military bases, installations, or fortifications and to abstain from testing any type of weapon or conducting military maneuvers on celestial bodies. The U.S. position identifies the Moon as a celestial body for purposes of this prohibition, but, again, that is arguably ambiguous. States Parties to the Treaty may verify treaty compliance on the Moon and other celestial bodies. Access requires reciprocity, advanced notice, consultations, safety precautions, and avoiding interference in the operation of a facility accessed. If two countries were at war, the U.S. position appears to be that the Treaty would not apply as between the two belligerents.

I. Basic Rules for Interpreting the Outer Space Treaty

A. A Treaty's Language Governs Its Interpretation

The Outer Space Treaty addresses a host of activities that a country may carry out in outer space and includes military activities amongst those it governs. Few if any court opinions address the Outer Space Treaty, which means that anyone seeking to interpret its meaning must rely on the text. Where the text is ambiguous, a reader may look to the treaty's drafting history to ascertain the intent behind unclear language.

This approach to interpretation is rooted in the law of the United States. For example, the first step relies on the rubric that the language of a treaty tells us what the treaty means. When interpreting a treaty, the Supreme Court explained that the treaty's unambiguous text governs its interpretation, regardless of the treaty's drafting history.[491] In the same case in which the Supreme Court applied this rule, it also pointed out that where the treaty language is clear, the court may not "insert an amendment."[492] In other words, a reader may not assume that the plain language of a treaty has somehow lost its effectiveness or contains a provision not present in writing because doing so might suit that reader's own purposes.

Only if a treaty provision is ambiguous will the courts consult the drafting history to help understand its meaning.[493] For example, although the court in *Narayanan* pointed out that the treaty's drafting history leant support to the court's interpretation, the court made sure to first disavow the need to even look at that history.[494]

Although the United States is not a signatory to it[495], the Vienna Convention on the Law of Treaties[496] ("Vienna Convention") offers

---

491. *Chan v Korean Air Lines*, 490 U.S. 122, 134 (1989); *Narayanan v British Airways*, 747 F3d 1125, 1131 (9th Cir. 2014).

492. *Chan v Korean Air Lines*, 490 U.S. 122, 134 (1989)(finding courts should read international agreements without "alter[ing], amend[ing], or add[ing] to any treaty, by inserting any clause, whether small or great, important or trivial"); *Narayanan v British Airways*, 747 F3d 1125, 1129 (9th Cir. 2014) ("Plaintiffs effectively ask us to write an implied fourth trigger into the Convention's terms. This we cannot do.").

493. *Chan v Korean Air Lines*, 490 U.S. 122, 134 (1989); *Narayanan v British Airways*, 747 F3d 1125, 1131 (9th Cir. 2014) (Where the language is unambiguous, a court need not assess a treaty's drafting history).

494. *Narayanan v British Airways*, 747 F3d 1125, 1131 (9th Cir. 2014) (Where the language is unambiguous, a court need not assess a treaty's drafting history).

495. Archives, Statement of U.S. Department of State, at https://2009-2017.state.gov/s/l/treaty/faqs/70139.htm (last checked Jan. 6, 2023).

496. Vienna Convention on the Law of Treaties; Vienna, done 23 May 1969, entered into force 27 January 1980; 1155 UNTS 331; UKTS 1980 No. 58; Cmnd. 4818; ATS 1974 No. 2; 8 ILM 679 (1969). Available at https://legal.un.org/ilc/texts/instruments/english/conventions/1_1_1969.pdf.

insights into how the international community interprets a treaty. The approach is very similar to that taken by the U.S. courts. Article 31, for example, provides that a treaty shall be interpreted in accordance with the ordinary meaning to be given its terms in their context and in the light of its object and purpose. Likewise, as U.S. courts allow, Article 32 permits a court to review preparatory work and the circumstances surrounding the treaty when the treaty terms are ambiguous or obscure.

There are other more detailed tools designed for idiosyncratic situations available for treaty interpretation, but the tools described here are those which must guide any analysis at the outset.

This review of the Outer Space Treaty's military requirements and prohibitions takes into account the testimony of the Treaty's negotiators to the U.S. Senate during the Treaty's ratification hearings. Testimony was provided by the chief U.S. treaty negotiator to the United Nations, the Secretaries of State and Defense, the Chairman of the Joint Chiefs of Staff, and the State Department's Legal Adviser. Consideration of their testimony serves two purposes. The Treaty does contain ambiguous or undefined terms, and the testimony helps explain the U.S. understanding of some of those terms. There are also points where the testimony appears to contradict the plain language of the Treaty, thus creating confusion for scholars in the decades that have followed.

This assessment will also highlight where the Outer Space Treaty contains ambiguities because it can be useful to know when the legal terrain may be shaky beneath one's feet. Even if the executive branch of the United States has been definite about an interpretation, that might not mean a court would agree. Additionally, another country might have an equally definite but opposite interpretation, and it is useful to know the possibility in advance, particularly if the other country is an adversary.

B. Customary International Law

Where a treaty does not address a disputed issue, where a treaty is silent, or where a treaty term is undefined, customary international law may apply. Customary international law—which is roughly akin to common law on the domestic front—contains a number of elements. The U.S. Department of State recently relied on the Statute of the International Court of Justice to identify them:

> To demonstrate the existence of a rule of customary international law, there must be "evidence of a general practice accepted as law," as stated in the Statute of the International Court of Justice (ICJ). This formulation reflects the two elements required for the formation of customary international law: (1) State practice and (2) opinio juris (acceptance as law).
>
> The formation of a rule of customary law requires that the relevant State practice be general and consistent; it must be "settled practice," as described by the ICJ. Moreover, such practice must have "occurred in such a way as to show a general recognition that a rule of law or legal obligation is involved" (opinio juris).
>
> For a rule to become part of customary international law, it is necessary for both the State practice and opinio juris requirements to be satisfied. Failure to satisfy either requirement has the effect that the suggested rule has not been established as a matter of international law.[497]

This analysis is useful when assessing whether a vague Outer Space Treaty term has acquired a definition under customary international law. This analysis will come into play in the context of determining

---

497. U.S. Dept. of State, *Limits in the Seas, No. 150, People's Republic of China: Maritime Claims in the South China Sea*, State Practice Supplement, 2 (Jan. 2022) (footnotes omitted). Available at https://www.state.gov/wp-content/uploads/2022/01/LIS150-SCS-Supplement.pdf. (Last checked Feb. 1, 2023).

whether the Treaty's prohibition on nuclear weapons in orbit around the Earth applies only to full orbits or to the abbreviated orbits of a fractional orbital bombardment system.

II. What Military Space Activities Does the Outer Space Treaty Prohibit?

Article IV of the Treaty contains a list of prohibitions on military space activity:

> *States Parties to the Treaty undertake not to place in orbit around the earth any objects carrying nuclear weapons or any other kinds of weapons of mass destruction, install such weapons on celestial bodies, or station such weapons in outer space in any other manner.*
>
> *The moon and other celestial bodies shall be used by all States Parties to the Treaty exclusively for peaceful purposes. The establishment of military bases, installations and fortifications, the testing of any type of weapons and the conduct of military manoeuvres on celestial bodies shall be forbidden. The use of military personnel for scientific research or for any other peaceful purposes shall not be prohibited. The use of any equipment or facility necessary for peaceful exploration of the moon and other celestial bodies shall also not be prohibited.*[498]

A. The Treaty Contains Restrictions on Nuclear Weapons and Any Other Kinds of Weapons of Mass Destruction.

The language of Article IV is plain with regard to nuclear weapons and any other kind of weapons of mass destruction ("WMD"). States

498. Treaty on Outer Space: Hearing before the Comm. on Foreign Relations of the U.S. Senate, 4, 90th Cong. (1967).

may not place such weapons in orbit around the Earth, install them on a celestial body, including the Moon, or station them anywhere else in outer space.

Nuclear weapons. The Outer Space Treaty does not define nuclear weapons, so official definitions must be found elsewhere. However, using principles of plain language, and the fact that nuclear weapons existed and had been used by 1967, when the Treaty was passed, we are likely safe in following a current definition provided by the U.S. Centers for Disease Control and Prevention (CDC), which describes a nuclear weapon as "a device that uses a nuclear reaction to create an explosion."[499] According to the CDC, nuclear weapons may come in the form of bombs or missiles.

Weapons of mass destruction. Although nuclear weaponry is a known technology and creates few ambiguities, the definition of a "weapon of mass destruction" is not as clear.[500] As with other terms, the Treaty does not define "weapons of mass destruction." However, to the extent the term is ambiguous, we may look to the Treaty's ratification history before the U.S. Senate to obtain guidance. Former Justice Arthur J. Goldberg, Ambassador to the United Nations, and the chief U.S. negotiator on the Outer Space Treaty explained during Senate ratification hearings that a weapon of mass destruction is "a weapon of comparable capability of annihilation to a nuclear weapon, biological."[501] U.S. Deputy Secretary of Defense Cyrus R. Vance's testimony clarified that chemical and biological weapons would be included in the definition.[502] Ambassador

---

499. Centers for Disease Control and Prevention, Environmental Health Infographics, "Nuclear Weapon." Available at https://www.cdc.gov/nceh/multimedia/infographics/nuclear_weapon.html (Last checked Jan. 19, 2023).
500. Lyall, F., and Larsen, P., *Space Law, A Treatise,* 2d ed., 460 (2018).
501. Treaty on Outer Space: Hearing before the Comm. on Foreign Relations of the U.S. Senate, 76, 90th Cong. (1967).
502. Treaty on Outer Space: Hearing before the Comm. on Foreign Relations of the U.S. Senate, 100, 90th Cong. (1967).

Goldberg further testified that observational or navigation satellites are not weapons of mass destruction.[503]

Lyall and Larsen appear not to consider Goldberg's guidance sufficiently clear. They view the Treaty's lack of definition for WMD as allowing, "considerable leeway for a narrow or wide interpretation."[504] They recommend looking to other sources of law[505] for making arguments in this regard. However, given the possible ambiguity surrounding the term, search for the U.S. position in this regard should start with the Ambassador's definition, and should focus, as he does, on whether a weapon's capacity for annihilation is on par with a nuclear weapon's capabilities.

Other weapons in orbit. According to the General Counsel of the U.S. Department of Defense (DOD), writing in the 2016 DOD Law of War Manual, the Treaty does not ban all weapons in orbit.[506] In fact, as DOD's Office of the General Counsel noted in 2002, "The treaty permits placing in orbit weapons other than nuclear weapons and other weapons of mass destruction."[507] The Treaty is silent with

---

503. Treaty on Outer Space: Hearing before the Comm. on Foreign Relations of the U.S. Senate, 77, 90th Cong. (1967).
504. Lyall, F., and Larsen, P., *Space Law, A Treatise,* 2d ed., 460 (2018).
505. Lyall, F., and Larsen, P., *Space Law, A Treatise,* 2d ed., 460 (2018).
506. General Counsel of the Department of Defense, U.S. *Department of Defense Law of War Manual*, 943 n. 161, Jun. 2015 (Updated Dec. 2016) (citing David A. Koplow, ASAT-isfaction: Customary International Law and the Regulation of Anti-Satellite Weapons, 30 M ICHIGAN JOURNAL OF INTERNATIONAL L AW 1187, 1198 (2009) ("This provision does not impede the stationing of non-nuclear weapons (including conventional ASAT weapons) in space, nor does it affect a nuclear weapon that makes only a temporary transit of outer space, as when propelled by an intercontinental ballistic missile (ICBM) toward its target, rather than being 'stationed' in space."). Available at https://dod.defense.gov/Portals/1/Documents/pubs/DoD%20Law%20of%20War%20Manual%20-%20June%202015%20Updated%20Dec%202016.pdf?ver=2016-12-13-172036-190. (Last checked Jan. 24, 2023).
507. Department of Defense, Office of the General Counsel, An Assessment of International Legal Issues in
Information Operations (2nd ed., Nov. 1999), reprinted and updated in 76 *U.S. Naval War College International Law Studies*, 459, 494 (2002). Available at https://digital-

respect to other types of weapons being stationed in orbit, and Russian state practice reflects at least one spacefaring nation's understanding of this interpretation, namely, that if the Treaty does not address other weapons, then it must not prohibit them. Historically and currently, Russian cosmonauts have carried guns to their orbiting platforms and to the International Space Station. The Russians considered them survival aids for after astronaut landings and before recovery in the Siberian wilderness, where bears and wolves were a concern.[508] They carried a triple barreled combination pistol from 1986 until 2006, and took a regular semi-automatic pistol from 2007, onwards.

More advanced weapons also fall outside the Article IV prohibition on nuclear and other WMD in orbit around the Earth. For space, those may include kinetic anti-satellite weapons which destroy an orbiting satellite through impact.[509] As Major Grunert[510] notes:

> Over the past year alone, Russia has tested both space-based and direct-ascent anti-satellite weapons — activities the United States has decried as "irresponsible," but which are not prohibited under existing international space law. China, which infamously destroyed one of its own satellites with a direct-ascent anti-satellite weapon in 2007, is also believed to be developing a wide range of anti-satellite and space weapons. The use of such weapons by China, Russia, or the United States would violate, or, at least, implicate, the Outer Space Treaty's Article III application of other areas of international law, including the laws of armed conflict, to

---

commons.usnwc.edu/cgi/viewcontent.cgi?article=1381&context=ils. (Last checked Feb. 1, 2023).

508. Wikipedia, TP-82, available at https://en.wikipedia.org/wiki/TP-82, (last checked Jan. 20, 2023).

509. Ferreira-Snyman, A. "SELECTED LEGAL CHALLENGES RELATING TO THE MILITARY USE OF OUTER SPACE, WITH SPECIFIC REFERENCE TO ARTICLE IV OF THE OUTER SPACE TREATY," Potchefstroom Electronic Law Journal (2015).

510. Major Jeremy Grunert is a U.S. Air Force JAG officer and attorney. He teaches space law as an assistant professor at the U.S. Air Force Academy.

> outer space. However, the Outer Space Treaty and international space law permit the development of such weapons and their placement into orbit, provided they do not constitute "weapons of mass destruction."[511]

Likewise, this provision of Article IV does not prohibit jamming and cyberattacks, which may render satellites inoperable. These activities may, however, be subject to the provisions of the Treaty's Article IX, which calls for States Parties to the Treaty to conduct their activities in the exploration and use of outer space "with due regard to the corresponding interests of all other States Parties to the Treaty."[512] Cyberattacks and jamming likely constitute the kind of harmful interference for which Article IX requires prior consultations. Likewise, the use of electro-magnetic pulse (EMP) weapons could run afoul of Article IX's requirement to consult prior to engaging in harmful interference. An EMP weapon when detonated,

> produces a pulse of energy that creates a powerful electromagnetic field capable of short-circuiting a wide range of electronic equipment, particularly computers, satellites, radios, radar receivers and even civilian traffic lights. Since EMP is electromagnetic energy traveling at the speed of light, all of the vulnerable electronic equipment in the detonation zone could be affected simultaneously.[513]

Were a nation to detonate an EMP in space without prior consultation with those who would be affected, the EMP's destruction of satellite capabilities would likely be deemed harmful interference under Article IX. Accordingly, even if a country placed

---

511. Grunert, J., "The 'Peaceful Use' of Outer Space?" *War on the Rocks*, (Jun. 22, 2021). Available at https://warontherocks.com/2021/06/outer-space-the-peaceful-use-of-a-warfighting-domain/ (Last checked Jan. 27, 2023).
512. The Outer Space Treaty, Art. IX.
513. Washington State Dept. of Health, "Electromagnetic Pulse," 2, Fact Sheet (Sept. 2003). Available at https://doh.wa.gov/sites/default/files/legacy/Documents/Pubs/320-090_elecpuls_fs.pdf (Last checked Mar. 30, 2023).

a nuclear weapon in only a "fraction" of an orbit so as to avoid the restrictions of Article IV as discussed below, the EMP could run afoul of Article IX.

Applicable orbits. To which orbits may Article IV's restriction on weapons of mass destruction, including nuclear weapons, apply? There is clarity regarding a full orbit and suborbital launches. There are theories and arguments to make about partial orbits, which are also known as fractional orbits.

Article IV's statement that "States Parties to the Treaty undertake not to place in orbit around the earth any objects carrying nuclear weapons or any other kinds of weapons of mass destruction," addresses the permissibility of such weapons orbiting the Earth, that is going around Earth in its entirety. For example, a nation-state might wish to place a nuclear weapon in low-Earth orbit, and just leave it there until needed. The nuclear weapon could orbit Earth for years, presumably to serve as a deterrent to poor behavior by potential adversaries. NASA defines an orbit as "a regular, repeating path that one object in space takes around another one."[514] If an orbit "around the Earth" consists of an object making a complete circuit then Article IV's prohibition on orbiting WMD applies, and the nation-state may not place a nuclear weapon in low-Earth orbit, and just leave it there until needed.

During the Senate ratification hearing for the Outer Space Treaty, the U.S. government's interpretation was clear: no signatory to the Treaty may place nuclear weapons or other WMD in orbit around the Earth under Article IV. Even if the Treaty itself were not so clear, a review of the preparatory work both reveals the official U.S. interpretation from the outset and confirms this interpretation. President Johnson welcomed the Treaty as "the most important arms

---

514. NASA, "What is an Orbit?" (Jul. 7, 2010). Available at https://www.nasa.gov/audience/forstudents/5-8/features/nasa-knows/what-is-orbit-58.html (Last checked Jan. 19, 2023).

control development since the limited Test Ban Treaty of 1963."[515] Ambassador Goldberg's testimony was even more pointed, referring repeatedly to the Treaty's prohibition on "bombs in orbit."[516] General Wheeler, Chairman of the Joint Chiefs of Staff, also spoke approvingly of the Treaty's "no bombs in orbit" provision.[517]

Might Article IV's prohibitions also include a suborbital trajectory? It does not appear so. Ambassador Goldberg's testimony addressed this issue clearly, although without much explanation of his rationale, in the context of whether Article IV applies to intercontinental ballistic missiles (ICBMs). We may deduce from the ambassador's testimony that Article IV's prohibition does not apply to ICBMs for two reasons. First, he explains that the Treaty does not apply to activities on the ground, stating that the Treaty does not cover ground activities at all. He then suggests that ICBMs must be dealt with in a separate treaty, not only because they launch from the ground but because they do not launch *into orbit*.[518] (Emphasis added). Given that an ICBM may reach an apogee of 2,000 kilometers (1,200 miles),[519] which far exceeds altitudes at which an object may achieve orbit, it is evident that Article IV's statement means what it says. As the DOD's Office of the General Counsel noted as of 2002, a nuclear weapon may pass through outer

---

515. Treaty on Outer Space: Hearing before the Comm. on Foreign Relations of the U.S. Senate, 115, 90th Cong. (1967).
516. Treaty on Outer Space: Hearing before the Comm. on Foreign Relations of the U.S. Senate, 22, 34, 90th Cong. (1967)(noting that Article IV puts into treaty form the "no bombs in orbit" resolution originally sponsored by the United States, and which was adopted by the U.N. [resolution] in 1963.
517. Treaty on Outer Space: Hearing before the Comm. on Foreign Relations of the U.S. Senate, 83, 89, 90th Cong. (1967).
518. Treaty on Outer Space: Hearing before the Comm. on Foreign Relations of the U.S. Senate, 72, 90th Cong. (1967) ("Ground activities are not covered at all by this treaty. We are dealing with activities in space, just like as I said earlier, that you have to leave for a disarmament treaty the question of ICBMs and their launching from ground and not into orbit.")
519. HandWiki, Encyclopedia of Knowledge, "Physics: Ballistic Missile," (undated). Available at https://handwiki.org/wiki/Physics:Ballistic_missile. (Last checked Jan. 21, 2023).

space but not be placed in orbit, and thus Article IV's restriction would not apply.[520]

While on the topic of ICBMs, it bears mention that Article IV also prohibits a State Party to the Treaty from stationing nuclear weapons and other kinds of WMD "in outer space in any other manner." This expansive language might seem at first blush to prohibit an ICBM's passage through outer space at all. However, Article IV prohibits states from undertaking to "station" such weapons in outer space in any other manner, indicating that such weapons may not be placed in outer space permanently or even, perhaps, for an extended duration. An ICBM may reach its intended destination in under an hour, while travelling far in excess of the speed of sound. This swift and limited travel creates no permanent "stationing" of the missile, and, thus, this portion of Article IV does not apply to ICBMs.

Nor do Lyall and Larsen consider ballistic missiles with suborbital trajectories prohibited by Article IV of the OST.[521] Their view is consistent not only with Goldberg's testimony, but with state practice subsequent to the Treaty's passage: the United States and the Russian governments have both possessed intercontinental ballistic missiles with nuclear warheads since then. Lyall and Larsen observe that although the trajectory of a ballistic missile may take it into space (assuming a clear boundary for "space" is agreed upon eventually), the missile would not be in orbit.[522]

A fraction of an orbit. Whether Article IV's WMD prohibition applies to complete orbits or suborbital trajectories is clear enough. It applies to the one but not the other. There is a more difficult and

---

520. Department of Defense, Office of the General Counsel, An Assessment of International Legal Issues in
Information Operations (2nd ed., Nov. 1999), reprinted and updated in 76 *U.S. Naval War College International Law Studies*, 459, 495 (2002). Available at https://digital-commons.usnwc.edu/cgi/viewcontent.cgi?article=1381&context=ils. (Last checked Feb. 1, 2023). Matte, N., *Aerospace Law*, 298 (1969).
521. Lyall, F., and Larsen, P., *Space Law, A Treatise,* 2d ed., 460, n.76 (2018).
522. Lyall, F., and Larsen, P., *Space Law, A Treatise,* 2d ed., 465 (2018).

ambiguous case, however: that of a fractional orbit, where a weapon "can stay on orbit as long as the user determines and then it de-orbits it as part of the flight path."[523] The Soviet Union's fractional orbital bombardment system was designed and tested in the early 1960s[524]—prior to the Outer Space Treaty's entry into force—to place a thermonuclear device in an incomplete orbit, one where it would not accomplish a full circuit of the Earth before diverting to its target. The Soviets claimed that Article IV's prohibition did not apply to an incomplete orbit, and went so far as to construct 18 operational FOBS silos.[525] More recently, the People's Republic of China has engaged in weapons test of a FOBS that employed a hypersonic glide vehicle.[526]

The U.S. Government appears to share the former Soviet Union's view that Article IV does not apply to fractional orbits. According to the U.S. DOD Law of War Manual:

> The prohibition on placing weapons of mass destruction "in orbit around the earth" refers only to their placement in full orbit around the Earth; thus, the Outer Space Treaty does not ban the use of nuclear or other weapons of mass destruction that go into a fractional orbit or engage in suborbital flight.158 For example, intercontinental ballistic missiles

---

523. Hitchens, T., "It's a FOBS, Space Force Saltzman confirms Amid Chinese Weapons Test Confusion," *BreakingDefense* (Nov. 29, 2021) (quoting Lt. Gen. Chance Saltzman, deputy Space Force chief for operations). Available at https://breakingdefense.com/2021/11/its-a-fobs-space-forces-saltzman-confirms-amid-chinese-weapons-test-confusion/ . (Last checked Jan. 24, 2023).
524. Eisel, B., "The FOBS of War," *Air Force Magazine*, 75 (Jun. 2005). Available at https://www.airandspaceforces.com/PDF/MagazineArchive/Documents/2005/June%202005/0605fobs.pdf (Last checked Jan. 24, 2023).
525. Eisel, B., "The FOBS of War," *Air Force Magazine*, 75 (Jun. 2005). Available at https://www.airandspaceforces.com/PDF/MagazineArchive/Documents/2005/June%202005/0605fobs.pdf (Last checked Jan. 24, 2023).
526. Hitchens, T., "It's a FOBS, Space Force Saltzman Confirms Amid Chinese Weapons Test Confusion," *BreakingDefense* (Nov. 29, 2021). Available at https://breakingdefense.com/2021/11/its-a-fobs-space-forces-saltzman-confirms-amid-chinese-weapons-test-confusion/ . (Last checked Jan. 24, 2023).

> (ICBMs) will travel a portion of their trajectory in outer space; but because ICBMs would enter outer space only temporarily, their entry into outer space with nuclear warheads would not violate this prohibition.159[527]

It is hard to argue with the interpretations of two major spacefaring nations; and although I could not find the Soviet rationale for its determination, the U.S. interpretation provides some explanation. The DOD Law of War Manual appears to rely on the Treaty's reference to an orbit "around the Earth," to mean "*all* around the Earth," not merely "*part way* around the Earth." However, neither limiting phrase "all" or "part way" are present in Article IV's text.

Despite these rather weighty interpretations to the contrary, it is evident that an ambiguity may exist regarding fractional orbits. Even when relying on the text, as the U.S. and Soviet positions purport to do, a different conclusion may be reached. Article IV does not say the Treaty's signatories undertake not to place restricted weapons in a *full* orbit. Instead, that provision is silent with respect to whether an orbit consists of a full or partial revolution around the Earth, thus indicating that the provision applies to any type of orbit around the Earth. That being so, the Treaty restriction could (and perhaps should) apply to any nuclear weapon or other WMD that achieves orbit, regardless of whether it achieves a full orbit or not. "Around the Earth" could refer to either a full orbit or to an orbit's location, that is, to contrast the orbit in question with one around the Moon, for example. Physicists speak of an object in orbit as "falling" around the Earth,[528] that is, in more colloquial terms, that it stays up there on its own. In contrast to an object achieving orbit, Congress has, as a matter of U.S. law, described a suborbital trajectory as

---

527. General Counsel of the Department of Defense, U.S. *Department of Defense Law of War Manual*. Jun. 2015 (Updated Dec. 2016). Available at https://dod.defense.gov/Portals/1/Documents/pubs/DoD%20Law%20of%20War%20Manual%20-%20June%202015%20Updated%20Dec%202016.pdf?ver=2016-12-13-172036-190. (Last checked Jan. 24, 2023).

528. Benson, T., "Free Falling Objects," NASA, updated May 13, 2021. Available at https://www.grc.nasa.gov/www/k-12/rocket/ffall.html.

meaning the intentional flight path of a launch vehicle, reentry vehicle, or any portion thereof, whose vacuum instantaneous impact point does not leave the surface of the Earth.[529] If a FOBS weapon's instantaneous impact point left the surface of the Earth, it would arguably no longer be on a suborbital trajectory and thus would be in orbit around the Earth, however briefly or temporarily. Both interpretations appear reasonable and grounded in the Treaty's text, and thus create ambiguity.

Scholars differ on whether Article IV prohibits carriage of a WMD in an incomplete orbit. For example, Lyall and Larsen maintain that an object's incomplete orbit counts as being in orbit for purposes of Article IV, and, thus, the Treaty prohibits the carriage on that object of a nuclear weapon or another weapon of mass destruction.[530] Other commenters see the matter differently,[531] with some saying that Article IV fails to address fractional orbital weapons at all.[532]

With it being possible to argue about the plain language of the prohibition, with spacefaring nations and scholars disagreeing, and with the fact that the former Soviet Union had started designing and testing its FOBS by 1965, two years before the Outer Space Treaty, it becomes evident we have a possible ambiguity here. Because of this, we may look to the restriction's underlying intent.

When the language of a law or treaty is not clear, a court may look at the intent underlying an ambiguous provision. Even that, however, is not helpful in this instance. There appears to be consensus that Article IV does not apply to ICBMs. Like an ICBM, a FOBS object is not in a permanent station around the Earth. They

529. 51 U.S.C. 50902(25). *See also* the FAA's implementing regulations at 14 C.F.R. 401.5.
530. Lyall, F., and Larsen, P., *Space Law, A Treatise,* 2d ed., 460, n.76 (2018) (asserting that fractional orbital bombardment systems are limited by Article IV).
531. See, e.g., Jasentuliyana, N. and Lee, R., *Manual on Space Law,* Vol. I, 13, 14, 254 (1979) (averring with neither explanation nor citation that the drafters did not intend the prohibition on WMD to apply to fractional orbits).
532. Strijdom, S.H, "The Use of Outer Space for Military Purposes, Article IV of the Outer Space Treaty," 35, Univ. of Pretoria (2013).

both merely use outer space to reach their targets. Because ICBMs are allowed under the Treaty, and because FOBS are functionally similar, perhaps the Treaty was not intended to apply to a system such as FOBS either. Moreover, it appears the drafters intended Article IV to stop nation-states from placing WMD in space on a permanent basis. An orbiting object may stay in space for years, even decades or more. The drafters clearly intended Article IV to prevent that situation. Likewise, Article IV prohibits "stationing" WMD in outer space in any other manner, again demonstrating the drafter's concern with nuclear weapons or other kinds of WMD being located in space for an extended duration. One might argue that the FOBS' limited duration passage through outer space does not violate the Treaty's intent.

One might also argue the contrary, namely, that a concern regarding permanence was not the only reason for prohibiting WMD in orbit. The ability to detect incoming nuclear warheads was also a concern. An ICBM launch is difficult enough to monitor and plan for. How much more difficult would it be to detect a warhead suddenly coming in from orbit, without benefit of first detecting the launch phase? If preventing surprise was the intent behind the Article IV prohibition, then a FOBS would not comply with that intent. More historical research on the question might prove fruitful.

Lastly, the FOBS scenario provides an opportunity to explore whether a WMD in a fractional orbit could be found compliant with Article IV as a matter of customary international law. The answer is not available without further historical research. Customary international law requires settled state practice and acceptance as law. Although one country, the Soviet Union, fielded a FOBS, it did not deploy it in its partial orbit (thankfully). Even though the construction of the silos might count as state practice, so might not launching them into a fractional orbit. The absence of a launch might be interpreted as an absence of state practice. Although the Chinese are testing a fractional orbital system, it is not one with a nuclear weapon or other kind of WMD, and thus provides no evidence of state practice with respect to such weapons. Research uncovered no

U.S. fractional orbiting of a nuclear weapon. Accordingly, the existence of settled state practice is unclear. More clear is the fact that both the former Russian government and the United States appear to view FOBS as compliant with Article IV, so that there is evidence of their acceptance of the interpretation as law. However, to find a rule of customary international law, both settled state practice and legal acceptance must be present. The former is not.

Installing WMD on the Moon. Article IV's restrictions on nuclear weapons and other weapons of mass destruction include a prohibition on installing such weapons "on celestial bodies." Legal scholar Nicholas Matte points out that the installation prohibition makes no mention of the Moon, which means that this restriction is ambiguous with respect to whether the Treaty would prevent a State Party from installing WMD on the Moon.[533] On the one hand, as Matte notes, the fact that the Treaty's title, other provisions, and a later portion of Article IV refer to celestial bodies as including the Moon suggests that the truncated reference implicitly includes the Moon, and that thus WMD may not be stationed there. On the other hand, Matte points out, under the rules of interpretation, the Moon's exclusion here when it is mentioned so many times in concert with other celestial bodies throughout the rest of the Treaty suggests that the exclusion was deliberate. Under the logic of the latter interpretation, nuclear weapons and other WMD could be stationed on Earth's Moon. However, the next provision is a catch-all that ensures that WMD may not be stationed in outer space in any other manner, and thus should resolve Matte's concerns.

Stationing WMD in outer space. Article IV also prohibits the Treaty's signatories from stationing WMD "in outer space in any other manner." This final prohibition on a country's ability to station nuclear weapons or other WMD "in outer space in any other manner" means that the prohibition is not limited to Earth's orbit and celestial bodies. Under the plain language of Article IV, the

533. Matte, N., *Aerospace Law*, 298 (1969).

prohibition applies to lunar orbit,[534] and, to orbits around any other celestial bodies. The DOD's Office of the General Counsel has been clear that such weapons may not be placed on the Moon,[535] and this particular prohibition could perhaps serve as the source of the Treaty's prohibition of nuclear weapons and WMD on the Moon.

Where outer space begins. Ambiguity surrounds the question of where outer space begins. Although the legal record provides clarity on whether certain locations are in airspace or outer space, the line where one becomes the other remains ambiguous. To date, the U.S. Government has declined to define the boundary between air space and outer space. There is a longstanding U.S. Government concern over identifying a boundary between air and space, dating back to the early 1960s. For example, the U.S. Air Force pointed out that establishing a "floor" for space might be premature because it could eventually be exceeded by high flying aircraft.[536] In 1962, the Joint Chiefs of Staff opposed prematurely defining outer space on the grounds that it could limit military space operations.[537] The Treaty itself establishes no definition, and Ambassador Goldberg's Senate testimony made it clear that any definition itself remained a subject for further and later study.[538]

---

534. Lyall, F., and Larsen, P., *Space Law, A Treatise,* 2d ed., 461 (2018).
535. Department of Defense, Office of the General Counsel, An Assessment of International Legal Issues in
Information Operations (2nd ed., Nov. 1999), reprinted and updated in 76 *U.S. Naval War College International Law Studies*, 459, 495 (2002)(" The Outer Space Treaty provides that the parties will not "place in orbit around the Earth any objects carrying nuclear weapons or any other kinds of weapons of mass destruction, install such weapons on celestial bodies [i.e., the Moon, planets, and asteroids], or station such weapons in outer space in any other manner.")(emphasis added). Available at https://digital-commons.usnwc.edu/cgi/viewcontent.cgi?article=1381&context=ils. (Last checked Feb. 1, 2023).
536. Terrill, D., *The Air Force Role in Developing Outer Space Law*, 56 (1999) *citing* DOD Planning Luncheon, paper, U.S. Initiative at the 16th General Assembly, (Sept. 12, 1961).
537. Terrill, D., *The Air Force Role in Developing Outer Space Law*, 57 (1999).
538. Treaty on Outer Space: Hearing before the Comm. on Foreign Relations of the U.S. Senate, 16, 90th Cong. (1967).

In addition to the official reasons for avoiding a definition of outer space, the lack of clarity provides other benefits. If, for example, one country sends a high-altitude vehicle over another country's territory, the sensed country has no basis for complaint if the vehicle is in outer space. This is because Article II of the Treaty provides that no country may claim sovereignty over outer space. Conversely, the sensed country does have grounds for complaint if the high-altitude vehicle enters its airspace without authorization. So long as no clear line demarcating the end of airspace and the beginning of outer space is identified, countries may address the question on a case-by-case basis. Technology may allow for vehicles to be operated at higher and higher altitudes. This may lead a country to seek to regulate more and more of its airspace. If a high-altitude vehicle could affect aircraft at lower altitudes or people on the ground, there might be strong reasons to regulate at these higher altitudes. We may not be able to currently foresee such technology, the uses to which it might be put, or the hazards it could cause, so preserving the ability to decide later could save on making a terrible mistake.

Nonetheless and despite the benefits of ambiguity, there are points of certainty. For example, high altitudes remain part of the air space of nation-states. Ambassador Goldberg testified during the Treaty's ratification hearings that the Treaty drafters had agreed that commercial aviation and supersonic air travel were not in outer space.[539] During the same hearings, Dean Rusk, the U.S. Secretary of State, reported a consensus, affirmed by Ambassador Goldberg, that an object in orbit is in outer space.[540] The Moon and other celestial bodies are in outer space, the Moon because it orbits the Earth, and other celestial bodies because they are beyond Earth orbit. Once we leave the murky area of where airspace ends and outer space begins, the Treaty makes clear that no nuclear weapons

539. Treaty on Outer Space: Hearing before the Comm. on Foreign Relations of the U.S. Senate, 28, 90th Cong. (1967).
540. Treaty on Outer Space: Hearing before the Comm. on Foreign Relations of the U.S. Senate, 17, 18, 90th Cong. (1967).

or other WMD may be installed on celestial bodies or stationed elsewhere in outer space, which should include cislunar space, lunar orbit, and, this writer suggests, the lunar surface.

B. States Parties Must Use the Moon and Other Celestial Bodies Exclusively for Peaceful Purposes. This Includes a Right of Self-defense.

As Secretary of State Dean Rusk noted, Article IV says that "Parties to the treaty undertake as well to use the moon and other celestial bodies exclusively for peaceful purposes." Two points merit mention. First, in applying this requirement only to the Moon and other celestial bodies, the provision avoids requiring that interplanetary space be exclusively peaceful. Second, scholars note that the United States views this provision as allowing the use of force when necessary for self-defense.

Interplanetary space need not be used exclusively for peaceful purposes. The requirement for States Parties to use the Moon and other celestial bodies exclusively for peaceful purposes, does not apply to interplanetary space.[541] For purposes of this discussion, "interplanetary space" is usage borrowed from Grunert, and is intended merely to signify areas that are not on the Moon or another planet. Interplanetary space may thus include "the areas between celestial bodies, including Earth orbit, cislunar space and cislunar orbit, and farther areas of deep space."[542] As Grunert succinctly phrases the matter, "In other words, while celestial bodies *are* to be used only for exclusively peaceful purposes, certain military uses of

541. Grunert, J., "The 'Peaceful Use' of Outer Space?" *War on the Rocks*, (Jun. 22, 2021). Available at https://warontherocks.com/2021/06/outer-space-the-peaceful-use-of-a-warfighting-domain/ (Last checked Jan. 27, 2023).
542. Grunert, J., "The 'Peaceful Use' of Outer Space?" *War on the Rocks*, (Jun. 22, 2021). Available at https://warontherocks.com/2021/06/outer-space-the-peaceful-use-of-a-warfighting-domain/ (Last checked Jan. 27, 2023).

interplanetary space are, by omission, *permitted*!"[543] In short, States Parties to the Treaty may use interplanetary space for non-peaceful purposes, and the Treaty doesn't stop them. The Treaty doesn't actually grant permission so much as not deny it.

Hence, we see military satellites and dual-use technology in orbit and otherwise in interplanetary outer space. Grunert observes that other treaties, including the failed Moon Treaty, have attempted to place "more expansive and explicit" restrictions on space weapons, but those attempts have not panned out.[544]

Self Defense is a peaceful use. Historically, while the U.S. maintained that peaceful purposes included the right to self-defense, the Soviet Union maintained that peaceful meant non-military.[545] As early as 1957, when Senator Albert Gore Sr. addressed a UN Committee about space, he stated that "it is the view of the United States that outer space should be used only for peaceful—that is, *non-aggressive and beneficial*—purposes..... Until [general disarmament] is achieved, the test of any space activity must not be whether it is military or non-military, but *whether or not it is consistent with the United Nations Charter and other obligations of*

---

543. Grunert, J., "The 'Peaceful Use' of Outer Space?" *War on the Rocks*, (Jun. 22, 2021). Available at https://warontherocks.com/2021/06/outer-space-the-peaceful-use-of-a-warfighting-domain/ (Last checked Jan. 27, 2023).

544. Grunert, J., "The 'Peaceful Use' of Outer Space?" *War on the Rocks*, (Jun. 22, 2021). Available at https://warontherocks.com/2021/06/outer-space-the-peaceful-use-of-a-warfighting-domain/ (Last checked Jan. 27, 2023).

545. Senator Albert Gore, Sr., Treaty on Outer Space: Hearings Before the Committee on Foreign Relations, U.S. Senate, 59, 90th Congress, (1967) ("We interpret peaceful purposes as being non-aggressive and beneficial."); *see also* General Counsel of the Department of Defense, U.S. *Department of Defense Law of War Manual*, 944 n.166 (suggesting linguistic differences in that "military" in the Russian language carries connotations of "warlike"), Jun. 2015 (Updated Dec. 2016). Available at https://dod.defense.gov/Portals/1/Documents/pubs/DoD%20Law%20of%20War%20Manual%20-%20June%202015%20Updated%20Dec%202016.pdf?ver=2016-12-13-172036-190. (Last checked Jan. 30, 2023).

law."[546] During the Treaty's Senate ratification hearings, Secretary of State Rusk noted that the UN Charter will apply in outer space,[547] and Article 51 of the UN Charter recognizes the right of self-defense. Attorneys within the Office of the Secretary of Defense and the services agreed that "peaceful purposes" were consistent with self-defense under the UN Charter.[548]

In the decades since the Treaty came into being, both the United States and the Russian governments have used outer space to orbit military satellites and conduct anti-satellite tests, thus providing evidence of state practice from both countries—as well as others—that "peaceful" need not mean non-military.

C. Military Bases, Installations, Fortifications, and Maneuvers Prohibited on Celestial Bodies

Lastly, in the words of Secretary of State Rusk during the Senate ratification hearings, Article IV says that "Parties to the treaty … undertake not to establish military bases, installations or fortifications and to abstain from testing any types of weapons or conducting military maneuvers on celestial bodies. There is, of course, no prohibition on the use of military personnel and equipment for peaceful purposes."[549]

The first question that arises is whether this provision applies to the Moon. A close reading shows that the Moon is omitted from the

---

546. Grunert, J., "The 'Peaceful Use' of Outer Space?" *War on the Rocks*, (Jun. 22, 2021)(quoting, with emphasis added, Sen. Gore's remarks to the U.N. General Assembly's First Committee, 15 (Dec. 3, 1962) available at https://www.unoosa.org/pdf/garecords/A_C1_PV1289E.pdf). Available at https://warontherocks.com/2021/06/outer-space-the-peaceful-use-of-a-warfighting-domain/ (Last checked Jan. 27, 2023).
547. Treaty on Outer Space: Hearing before the Comm. on Foreign Relations of the U.S. Senate, 4, 90th Cong. (1967).
548. Terrill, D., *The Air Force Role in Developing Outer Space Law*, 57 (1999).
549. Treaty on Outer Space: Hearing before the Comm. on Foreign Relations of the U.S. Senate, 4, 90th Cong. (1967).

prohibition. Following Professor Matte's earlier logic discussed above, he struggled with the question of whether the omission means that the Treaty prohibits military bases, installations, fortification, testing, or maneuvers on the Moon, or only on other celestial bodies. Under common rules of interpretation, the Moon's exclusion here—when it is mentioned so many times in concert with other celestial bodies throughout the rest of the Treaty—suggests that the exclusion was deliberate.[550] When the Treaty's drafters meant a provision to apply to the Moon, they said so repeatedly. Here they did not. As the Supreme Court noted in *Chan v. Korean Airlines*, in interpreting a treaty, "courts should read international agreements without "alter[ing], amend[ing], or add[ing] to any treaty, by inserting any clause, whether small or great, important or trivial."[551]

The DOD Law of War Manual suggests Professor Matte should not have been concerned. The Manual states that "Article IV of the Outer Space Treaty places certain prohibitions on military activities *on the moon* and other celestial bodies: (1) the establishment of military bases, installations, and fortifications; and (2) the testing of any type of weapons; and (3) the conduct of military maneuvers."[552] (Emphasis added). On its face, the Manual appears to include the Moon as being a location where these activities are prohibited. Implicit in the Manual's assertion is the apparent conclusion that celestial bodies always include Earth's Moon. However, the Manual does not address the legal concern occasioned by the Moon's absence from the text, and thus provides no information on how the conclusion was reached. In support of the Manual's conclusion,

---

550. Matte, N., *Aerospace Law*, 299 (1969).
551. *Chan v Korean Air Lines*, 490 U.S. 122, 134 (1989).
552. General Counsel of the Department of Defense, U.S. *Department of Defense Law of War Manual*, 941 n.153, Jun. 2015 (Updated Dec. 2016). Available at https://dod.defense.gov/Portals/1/Documents/pubs/DoD%20Law%20of%20War%20Manual%20-%20June%202015%20Updated%20Dec%202016.pdf?ver=2016-12-13-172036-190. (Last checked Feb. 1, 2023).

when referring to using "[t]he moon and other celestial bodies" exclusively for peaceful purposes, Article IV's reference to "other celestial bodies" suggests that the Moon is indeed a celestial body. This reference might be a permissible basis for the Manual's implicit conclusion.[553]

To the extent the provision itself is ambiguous, more research into the provision's historical intent may be required because it would be hard to claim intent is clear without that information. The Treaty was drafted before the first man had even set foot on the Moon. Different intents may be reasonable to surmise. On the one hand, it would make sense for someone to want to preserve the Moon for military bases and maneuvers, implying that the Moon's omission was deliberate. After all, the Treaty does not prohibit nuclear ICBMs. On the other hand, the Moon's proximity to Earth would provide a good reason to attempt to keep it free of military activity, and we might rely on the reference to the Moon as an "other celestial body" elsewhere in Article IV to determine that the prohibitions on establishing military bases, installations or fortifications and on testing any types of weapons or conducting military maneuvers indeed apply to the Moon.

D. Article XII Allows for Inspections with Reciprocity and Advance Notice.

The Treaty's Article XII provides a means of verifying Treaty compliance:

> All stations, installations, equipment and space vehicles *on the moon and other celestial bodies* shall be open to

553. *See also*, Department of Defense, Office of the General Counsel, "An Assessment of International Legal Issues in Information Operations." (2nd ed., Nov. 1999), reprinted and updated in 76 *U.S. Naval War College International Law Studies*, 459, 495 (2002). Available at https://digital-commons.usnwc.edu/cgi/viewcontent.cgi?article=1381&context=ils. (Last checked Feb. 1, 2023).

> representatives of other States Parties to the Treaty on a basis of reciprocity. Such representatives shall give reasonable advance notice of a projected visit, in order that appropriate consultations may be held and that maximum precautions may be taken to assure safety and to avoid interference with normal operations in the facility to be visited. (Emphasis added) [554]

As is evident from the text, the Treaty limits inspections to the Moon and other celestial bodies. For example, this Article does not apply to stations, installations, equipment, or vehicles on the ground on Earth.[555] As Ambassador Goldberg noted in his Senate testimony, the Soviets held up negotiations for two months when they tried to insist on access to U.S. tracking facilities anywhere in the world, but the United States did not agree.[556] Additionally, the Chairman of the Joint Chiefs, General Wheeler, noted in his Senate testimony that this provision did not apply to facilities on orbit.[557] Thus, one nation would not be able to access another country's orbiting satellite.

One area of ambiguity may be found in the requirement for reciprocity. Article XII provides that facilities on celestial bodies shall be open "on a basis of reciprocity." On its face, this language could be understood to mean that countries have to agree with each other on whether to allow inspections. A court might, in fact, look no further if it had to interpret the question.

Ambassador Goldberg, however, explained that the reciprocity requirement does not grant a nation a "veto" power of another country's right of access. The Ambassador explained that the Soviets

---

554. Wheeler, Chairman, Joint Chiefs, Treaty on Outer Space: Hearing before the Comm. on Foreign Relations of the U.S. Senate, 84, 90th Cong. (1967); *See also id.,* Goldberg at 115, Meeker at 101-102.
555. Treaty on Outer Space: Hearing before the Comm. on Foreign Relations of the U.S. Senate, 77, 90th Cong. (1967).
556. Treaty on Outer Space: Hearing before the Comm. on Foreign Relations of the U.S. Senate, 77, 90th Cong. (1967).
557. Treaty on Outer Space: Hearing before the Comm. on Foreign Relations of the U.S. Senate, 91, 90th Cong. (1967). *See also id.*, Goldberg at 19.

had insisted on the reciprocity, but the United States did not want the language to suggest it allowed anyone to deny access under Article XII. Accordingly, Ambassador Goldberg made sure the UN record reflected the U.S. understanding and its reliance on that understanding, making an official, on-the-record statement at both the UN's Legal Subcommittee and the Political Committee of the General Assembly.[558] He then explained the substance of the matter as follows:

> In these clarifying statements [at the UN] I pointed out that the words "on the basis of reciprocity" in Article XII do *not* import a veto. That is, they do *not* mean that State A may visit State B's facilities or installations on a celestial body only if B asks to visit those of A. On the contrary, "on the basis of reciprocity" merely states what would be true in any event under international law. Any party to the Treaty has the right to visit installations of another party on a celestial body—whether or not the other party chooses to exercise its reciprocal right. If, however, the prospective visitor has illegally and in violation of the treaty, barred visits to its facilities by the State whose installations it wishes to inspect, the second state may deny a visit to the breaching party. This result is simply an application of the principle that when one party breaches a material obligation which is owed to another party, the latter is entitled to withhold performance of a commensurate obligation which it would otherwise have owed to the first party. ... Thus, in the event of a material breach, a party would have the option of treating the entirety of its treaty obligations toward the breaching party as having come to an end, to be revived only upon remedial action by the defaulter.[559]

558. Treaty on Outer Space: Hearing before the Comm. on Foreign Relations of the U.S. Senate, 152, 90th Cong. (1967).
559. Treaty on Outer Space: Hearing before the Comm. on Foreign Relations of the U.S. Senate, 152, 90th Cong. (1967).

In short, this provision of the Treaty requires a country of whom access is requested to agree to it. The requesting country must also agree to any request. No preliminary reciprocity agreements are required. The Treaty does, however, require advance notice of a projected visit for purposes of taking safety precautions and to avoid interference with normal operations.[560] The Moon's precarious circumstances with respect to oxygen and other necessities mandate planning ahead.[561] Additionally, any verification activities must themselves comply with Article IX's admonition against interfering with other states.

E. Whether the Treaty Applies During War Is Unlikely as between Belligerents.

The Outer Space Treaty's applicability provision at Article XIII states that "The provisions of this Treaty shall apply to the activities of States Parties to the Treaty in the exploration and use of outer space, including the Moon and other celestial bodies, ..." If the Treaty applics only to "exploration and use" it may be argued that neither of those activities includes war, although a broader interpretation might view war as a "use." However, General Wheeler testified to the Senate during the 1967 ratification hearings that the Treaty would *not* apply during war.[562] DOD's Office of the General Counsel later provided a fuller explanation in 1999 and as updated in 2002:

> During an international armed conflict between the two nations concerned, however, the law of armed conflict would apply unless it was trumped by the principle of noninterference with space systems. Resolution of this issue

560. Treaty on Outer Space: Hearing before the Comm. on Foreign Relations of the U.S. Senate, 44, 90th Cong. (1967).
561. Treaty on Outer Space: Hearing before the Comm. on Foreign Relations of the U.S. Senate, 44, 90th Cong. (1967).
562. Treaty on Outer Space: Hearing before the Comm. on Foreign Relations of the U.S. Senate, 98, 90th Cong. (1967).

> depends largely on whether the four space treaties will be considered to apply during an armed conflict. None of them has any specific provision that indicates whether the parties intended that the agreement apply in wartime. There appears to be a strong argument that the principle of noninterference established by these agreements is inconsistent with a state of hostilities, at least where the systems concerned are of such high military value that there is a strong military imperative for the adversary to be free to interfere with them, even to the extent of destroying the satellites in the system. As indicated in the discussion of treaty law in the introduction to this paper, the outcome of this debate may depend on the circumstances in which it first arises in practice. Nevertheless, it seems most likely that these agreements will be considered to be suspended between the belligerents for the duration of any armed conflict, as least to the extent necessary for the conduct of the conflict.[563]

In short, the U.S. Department of Defense seems unlikely to apply the strictures of the Outer Space Treaty during an armed conflict. The DOD position does not appear to have changed as of the 2016 DOD Law of War Manual, which states in reliance on the 1999 assessment quoted above: "Certain provisions of these treaties may not be applicable as between belligerents during international armed conflict."[564]

---

563. Department of Defense, Office of the General Counsel, An Assessment of International Legal Issues in
Information Operations (2nd ed., Nov. 1999), reprinted and updated in 76 *U.S. Naval War College International Law Studies*, 459, 494 (2002). Available at https://digital-commons.usnwc.edu/cgi/viewcontent.cgi?article=1381&context=ils. (Last checked Feb. 1, 2023).

564. General Counsel of the Department of Defense, U.S. *Department of Defense Law of War Manual*, 941 n.153, Jun. 2015 (Updated Dec. 2016). Available at https://dod.defense.gov/Portals/1/Documents/pubs/DoD%20Law%20of%20War%20Manual%20-

PART IV

%20June%202015%20Updated%20Dec%202016.pdf?ver=2016-12-13-172036-190. (Last checked Feb. 1, 2023).

China Waging War in Space: An After-Action Report

August 2021

Executive Summary

This debrief reflects not only what happened during a three-move wargame NPEC hosted in June, but the game's three preparatory meetings. Game participants represented the United States, the People's Republic of China (PRC), and Japan and U.S. Indo-Pacific allies. The first move was set in 2027. The second and third moves were set in 2029. In the first move, PRC initiated space control operations intended to weaken U.S. regional allied (e.g., Japan, Republic of Korea (RoK), Australia) resolve in opposing a PRC economic exclusion zone enforcement effort directed against Taiwan.

The game produced four takeaways:

1. ***The United States and its space allies will have to defend against new, immediate Chinese co-orbital anti-satellite threats and make efforts to operate near and beyond the moon, or risk losing free access to space.*** Chinese co-orbital anti-satellite weapons capabilities (ASATs) in low Earth orbit (LEO) and Geosynchronous orbits (GEO) are real and growing and require immediate U.S. and allied acquisition and pre-positioning of bodyguard spacecraft to protect high-value military satellites. Pre-positioning is essential given that orbital mechanics do not allow for last-minute launches or maneuvers. Unfortunately, neither the United States nor its

space allies have bodyguard spacecraft on station with the maneuverability required to protect their military satellites. Chinese efforts to control cislunar space are also a major emerging threat that the U.S. and its allies' moon exploratory efforts do not adequately address. Unless America and its space allies act more quickly to address these issues, China will get the right stuff at the right time and gain much more control of space than America and its allies.

2. ***The U.S. and its space allies must give greater attention to how commercial space systems could be targeted and used for military purposes***. Peaceful space operations – co-orbital satellite servicing and refueling, debris removal, laser satellite tracking, etc. – could be quickly flipped from legitimate civil activities to hostile military actions. At a minimum, the U.S. and its space allies should be worried that China might use its financial clout to buy and control foreign commercial space firms (especially those operating in states that are not signatories to the Outer Space Treaty). This could allow China to use these firms' space systems to undermine U.S. and allied space operations with plausible deniability.

3. ***Given the history of Russian and Chinese gaming of diplomatic agreements, and the ambiguities associated with space controls, the U.S. and its space allies should reach no new space agreements with either China or Russia unless those understandings are clearly enforceable***. This includes agreements to clarify redlines and penalties for when the redlines are crossed. Because hostile space operations can produce significant strategic military results quickly, any effective space agreement should give all parties the unilateral right to protect their own assets and place the burden of proof on the accused party. Finally, enforcement actions should be proportional. The scope of defensive

actions should be limited to measures sufficient to bring the violating party back into compliance; no more, no less. For example, bodyguard spacecraft should be used not to destroy hostile space systems, but to gently push offending satellites far enough away that they no longer violate agreed safety zones. Also, certain laser, cyber and electronic warfare systems can be operated only to disable their targets temporarily. Agreements that meet the criteria noted (agreed to either internationally or, at least, among America's space allies) would make it far easier to orchestrate timely, effective responses to hostile space actions. Agreeing to anything less, however, could risk encouraging just the opposite.

4. ***The United States will have no chance of contesting or besting China in space unless Washington works much more closely with its space allies.*** Being able to call on NATO, the EU and America's Indo Pacific allies to deescalate or prevail in a space confrontation is an advantage the United States enjoys over China. America and its space allies must maintain this advantage by increasing U.S.-allied space training, planning, and gaming. America's space allies need to know in advance what the United States is likely to do in a variety of space conflict scenarios. Any major surprises on this front will come at a steep cost to U.S.- space allied relations. Finally, the game demonstrated that it is far less likely that China can successfully peel Japan from the United States than it might separate South Korea from the United States. This recommends increasing U.S. and allied space-related cooperation with Seoul.

Game Format and Scenario

Game participants were assigned to one of three teams: (1) United States, (2) People's Republic of China, and (3) Japan and Regional Allies (Australia, RoK, Taiwan and other nations in the region). The control center observed and conducted overall game administration (the team assignments and participant's roles are available here).

Move one began in 2027 and participants engaged in operational space capabilities that were assumed to be in an experimental phase. Each team was provided a set of capabilities with which they could conduct space situational awareness, and offensive and defensive space control operations. Commercial assets resided in all orbits by 2027-2029, and teams were able to target these assets as desired or use these assets to support military missions.

Move two replayed move one in 2029 with each team having more space operational capabilities. Finally, move three examined the long-term implications of the crisis. After move three, teams were asked two questions: What worked and what didn't and what new initiatives and capabilities might have made a decisive difference in the game's outcome.

At the start of move one (2027), the PRC had not put significant naval, air, or ground forces behind its enforcement of its declared Taiwan exclusive economic zone. The PRC's goal in implementing the Taiwan exclusive economic zone was not to starve the island, but to increase China's political jurisdiction and control over its renegade province.

In NPEC's earlier Taiwan wargame, held in October of 2020, China attempted military enforcement of an exclusive Taiwan economic zone. This devolved into a series of air and naval battles with the United States and Japan. These battles resulted in significant casualties and loss of assets for all sides and left the long-term resolution of the conflict very much in doubt.

In this China space wargame, the PRC took a different course, it initiated space control operations in hopes of weakening U.S. resolve and separating American regional allies (e.g., Japan, RoK, Australia) from opposing China's future economic zone enforcement effort against Taiwan. For this scenario (see space orders of battle), the PRC had ten rendezvous satellites that could serve as ASATS, and up to 17 ground-based laser systems that might be used for ASAT purposes. The 10 service satellites' commercial raison d'etre was to rendezvous with an operational satellite to provide refueling. However, the satellites' robotic arms also could be used to interfere with another satellite without its permission, essentially acting as an ASAT platform. In the American case, ASAT capabilities included several first-generation commercial re-fueling satellites in GEO, two commercial satellites with "space-tug" capability, and the U.S. X-37B space plane for possible LEO anti-satellite missions, and several commercial laser-track systems, (such as the EOS system in Australia) with enough power to move LEO space debris from collisions with active satellites.

Move One - 2027

The PRC team deployed ten rendezvous ASAT systems into orbits within reach of the U.S. nuclear command, control, and communications space infrastructure. The aim here was to deter the United States while preserving some degree of plausible deniability.

Several PRC players thought China should have held some of its rendezvous ASATs in reserve while threatening critical Japanese and U.S. Position-Navigation-Timing (colloquially referred to as GPS) satellites. They thought this would be less provocative and so, less risky. A suggested alternative to attacking allied GPS satellites was to jam their coverage over Taiwan. These launch and deployment decisions were supported by the operational deployment of PRC mobile and fixed ASAT laser platforms to threaten U.S. intelligence satellites in LEO.

As the PRC team waited for the United States to respond, it debated China's next move. The team agreed that China's primary goal was to create tension between the United States and its key Indo-Pacific allies. Actions that created economic disincentives for Japan and America's other regional allies were discussed at length. The approaches the China team considered included lazing or jamming Japanese satellites, creating small disruptions in key services every time a U.S. warship entered allied waters; using dual-use or commercial assets to take overt action while claiming the private operators acted of their own accord; and seeding doubt among allies by damaging the PRC's own satellites and claiming that Australia or Singapore was responsible.

Instead of any of these options, the China team decided to threaten specific US nuclear command control and communications satellites that were essential to deliver on America's nuclear guarantees to its allies. This drew a harsh demarche from the United States and the repurposing and repositioning of several U.S. satellites as "escorts" for the threatened nuclear command, control, and communications satellites. The United States also deployed the X-37B space plane, which was able to put comparable PRC nuclear arsenal command, control, and communications satellites at risk.

The Japan and Regional Allies team thought the PRC efforts to threaten U.S. nuclear command and control satellites was an aggressive warning and were eager to learn what Washington's response would be, as well as what diplomatic options were available. The United States wanted first to consult with its space allies and then move to the United Nations to present a clear view of what constitutes a space safety zone and to clarify what self-defense actions could be taken in response to potentially hostile operations within these zones against a nation's space assets.

After this first move, the participants agreed that:

1. The United States and its space allies would benefit from an international agreement that clarified what constitutes irresponsible and aggressive behavior in space. Although the odds of the PRC ever agreeing to such an understanding seem slim, Washington will have real difficulty leading any alliance space efforts if it fails to seek such an understanding, at least among its allies.

2. Japan, South Korea, Australia, and other US space allies are more easily allied with the United States than they are with each other. All would likely look to the United States for assistance in response to hostile PRC space actions. PRC threats or actions against their commercial or sovereign assets, or a regional denial of GPS capabilities would also prompt them to seek U.S. support and assistance.

3. The type of assistance America's space allies would seek from the United States would include information sharing, a commitment to defend allied satellites, access to U.S. space capabilities to compensate for any loss of space capabilities they themselves might experience, and a U.S. commitment to use offensive nuclear and conventional force to deter China.

4. The PRC might invest in or otherwise gain control of third-party commercial space service companies and use those assets in gray, deniable operations to provoke or embarrass the United States and its allies.

5. Rendezvous satellites and other dual-use technologies (e.g., ground-based lasers) are likely to become more developed and ubiquitous requiring international scrutiny and space situational awareness.

6. The United States should encourage the rapid launch of satellite constellations into LEO to replace large, vulnerable LEO systems.

7. The United States should accelerate the development of small satellite technologies to disperse and create greater resiliency in critical space functions (such as nuclear command and control) while at the same time creating a cheap proliferated class of "escort" satellites that can run interference for large, expensive legacy systems.

8. The United States should acquire a fleet of space planes or spacecraft that have enough fuel onboard to easily move to different orbits, quickly. This fleet can and should include both military and commercial systems.

Move Two – The First Six Months of 2029

Move two was a restart of the 2027 scenario in 2029. The primary purpose for replaying move one was to allow the teams to rethink their earlier diplomatic and military strategies with a more robust space order of battle.

At the start of move two, the United States tried unsuccessfully to establish international guidelines to define aggressive space behavior. Washington proposed a summit of the major space powers but the PRC demanded that the proposed meeting be moved to a UN venue that would force the United States to seek consensus. Once the UN meeting was convened, the Chinese insisted that historical

U.S. “inspection” actions were no different than what the PRC proposed to do with its own rendezvous service satellites and vetoed America’s proposals. This blocked agreement.

The PRC, though, lacked true allies. Instead, it only had client states. Washington, in contrast, could count on several Asian nations, including Japan, RoK, Taiwan, and Australia, to stand with it. In addition to its Asian allies, the United States expected some level of European Union and NATO support against aggressive PRC activities in space. The Japan and Regional Allies team saw strengthening the “Quad” (the United States, Japan, Australia, and India) as an additional counterweight to Chinese expansionism.

In an effort to undermine Washington’s space alliance efforts, the PRC economically entangled various Asian nations, offering them access to Chinese lunar facilities and China’s space station (assumed to exist in some form by 2029). China reached out to countries Washington had excluded from NASA’s Artemis Program. The U.S. and Japan and Regional Allies teams viewed these PRC efforts as an extension of China’s One Belt One Road initiative, i.e., as a further extension of the PRC’s client state strategy. The Japan and Regional Allies team, though, thought these initiatives were unlikely to neutralize Japan, which would draw closer to the United States if threatened by the PRC.

The crisis early in 2029 begins when the PRC secretly starts launching ten rendezvous satellites into various orbits close to U.S. military satellites. The Chinese then augmented this effort with periodic ground-based jamming of Japanese naval communications satellites and the lazing of Japanese Earth observation satellites.

As the PRC launches progressed over the first six months of 2029, the United States allowed the Japanese to use America’s military communications satellites. The United States also launched “defensive” escort satellites that could prevent PRC rendezvous satellites from freely threatening the Japanese satellites. The United States complimented these efforts by sharing intelligence, reconnaissance, and surveillance information with Japan, Australia, RoK, and Taiwan. The United States attempted to draw the EU into the crisis by offering EU nations use of U.S. GPS assets should future PRC activities degrade EU assets.

The Japan and Regional Allies team noted that most allied nations wanted to secure a diplomatic solution to the crisis. However, Australia was also willing to make reversible offensive strikes, with cyber and electronic warfare actions, or pre-emptive self-defense. In addition, the U.K. and France were willing to employ their bodyguard satellites in support of the United States (so long as their operation inflicted no irreversible harm), but German hesitancy blocked EU and NATO consensus to support such action. The Japan and Regional Allies team noted that the United States would have difficulty enlisting the support of the RoK unless the RoK thought lending such support would improve its defenses against North Korea's nuclear missiles.

This was the situation at the close of the sixth month in the 2029 space crisis.

Move Three – The Second Six Months of 2029

The seventh month in the 2029 space crisis began with a Japanese radar Earth observation satellite in LEO falling into a deteriorating orbit. The cause was unknown. The United States and Japan and Regional Allies teams tried to ascertain what caused the satellite's rapidly deteriorating orbit and the inability to recover it using ground control stations. Unfortunately, U.S. and Japanese space situational awareness capabilities were inadequate to shed light on this.

This prompted a discussion, which focused on four possible U.S. – allied responses:

The first was diplomatic. The United States team called for an immediate summit to define what constituted "defense zones," which ought not be violated without permission of the controlling nation. The Japan and Regional Allies team stipulated that rules regarding defense zones would be more effective if they specified the consequences of violating the zones.

The PRC categorically rejected any such zoning rules arguing that "they are in blatant violation of Article II of the Outer Space Treaty that outer space 'is not subject to national appropriation by claim of sovereignty, by means of use or occupation, or by any other means." China went on to note that it would order its satellites to exercise the "right of freedom of navigation" into any number of these zones and would retaliate if its right is infringed. It also noted that the United States had previously inspected China's satellites and that China was free to do the same.

The second possible response was using reversible space countermeasures. In this regard, most of the allied players favored using cyberweapons and low-powered lasers (for dazzling) to interfere with Chinese military and rendezvous satellites. These techniques do not create debris and do no lasting harm.

The third was deploying "bodyguard" or escort spacecraft to protect critical Japanese and U.S. satellites. These bodyguard spacecraft would not destroy Chinese satellites or spacecraft, but only push them away from key U.S. and allied satellites or temporarily disable them. The low inventory of U.S. bodyguard satellites in the 2029 scenario limited what the United States and its allies could do. This shortfall was shored up to a limited extent with the X-37B space plane, which could make close approaches to PRC rendezvous systems in LEO.

The fourth possible response was the sharing of U.S. space situational awareness information, intelligence surveillance reconnaissance data, and communications (particularly Naval communications) to compensate for PRC interference with allied satellites. The sharing of these capabilities strengthened U.S. alliance ties.

Finally, the U.S. and Japan and Regional Allies teams discussed taking military action against terrestrial targets, but ultimately did not agree on what specific targets to hit given fears of escalation. That noted, both the PRC and Japan and Regional Allies teams were skeptical that moves in space alone would enable the PRC to break U.S. allies away before enforcing a Taiwan economic exclusion zone.

Despite American efforts to work all of the agreed categories of responses, the PRC escalated its space operations. In addition to the 60 rendezvous spacecrafts the PRC had already placed in various orbits, China continued to launch, in fast pace, the remaining 48 of the 108 rendezvous spacecraft (100 from PRC space order of battle plus eight repurposed from 15 such spacecraft originally intended for satellite servicing and space debris removal). These 108 spacecraft were prepositioned in the vicinity of all U.S. (86), Japanese (9), Australian (7), South Korean (6) and Taiwanese (0) GEO, MEO and HEO satellites (as shown in their space orders of battle respectively). Also, Chinese ground-based lasers could damage LEO satellite sensors and even exterior structures.

Wargame Takeaways

This game drove home four important points.

1. ***The United States and its space allies will have to defend against new, immediate Chinese co-orbital anti-satellite***

threats and make efforts to operate near and beyond the moon, or risk losing free access to space. Chinese co-orbital anti-satellite weapons capabilities (ASATs) in low Earth orbit (LEO) and Geosynchronous orbits (GEO) are real and growing and require immediate U.S. and allied acquisition and pre-positioning of bodyguard spacecraft to protect high-value military satellites. Pre-positioning is essential given that orbital mechanics do not allow for last-minute launches or maneuvers. Unfortunately, neither the United States nor its space allies have bodyguard spacecraft on station with the maneuverability required to protect their military satellites. Chinese efforts to control cislunar space are also a major emerging threat that U.S. and its allies' moon exploratory efforts do not adequately address. Unless America and its space allies act more quickly to address these issues, China will get the right stuff at the right time and gain much more control of space than America and its allies.

2. ***The U.S. and its space allies must give greater attention to how commercial space systems could be targeted and used for military purposes***. Peaceful space operations – co-orbital satellite servicing and refueling, debris removal, laser satellite tracking, etc. – could be quickly flipped from legitimate civil activities to hostile military actions. At a minimum, the U.S. and its space allies should be worried that China might use its financial clout to buy and control foreign commercial space firms (especially those operating in states that are not signatories to the Outer Space Treaty). This could allow China to use these firms' space systems to undermine U.S. and allied space operations with plausible deniability.

3. ***Given the history of Russian and Chinese gaming of diplomatic agreements, and the ambiguities associated with space controls, the U.S. and its space allies should reach no***

new space agreements with either China or Russia unless those understandings are clearly enforceable. This includes agreements to clarify redlines and penalties for when the redlines are crossed. Because hostile space operations can produce significant strategic military results quickly, any effective space agreement should give all parties the unilateral right to protect their own assets and place the burden of proof on the accused party. Finally, enforcement actions should be proportional. The scope of defensive actions should be limited to measures sufficient to bring the violating party back into compliance; no more, no less. For example, bodyguard spacecraft should be used not to destroy hostile space systems, but to gently push offending satellites far enough away that they no longer violate agreed safety zones. Also, certain laser, cyber and electronic warfare systems can be operated only to disable their targets temporarily. Agreements that meet the criteria noted (agreed to either internationally or, at least, among America's space allies) would make it far easier to orchestrate timely, effective responses to hostile space actions. Agreeing to anything less, however, could risk encouraging just the opposite.

4. ***The United States will have no chance of contesting or besting China in space unless Washington works much more closely with its space allies.*** Being able to call on NATO, the EU and America's Indo-Pacific allies to deescalate or prevail in a space confrontation is an advantage the United States enjoys over China. America and its space allies must maintain this advantage by increasing U.S.-allied space training, planning, and gaming. America's space allies need to know in advance what the United States is likely to do in a variety of space conflict scenarios. Any major surprises on this front will come at a steep cost to U.S.- space allied relations. Finally, the game demonstrated that it is far

less likely that China can successfully peel Japan from the United States than it might separate South Korea from the United States. This recommends increasing U.S. and allied space-related cooperation with Seoul.

Pyongyang Goes Nuclear in Space: An After-Action Report

January 2023

Wargame Scenario and Rationale

In the early 2000s, many hoped states might rely less on nuclear weapons for their security. Today, that hope seems remote. China is dramatically increasing the size and quality of its nuclear arsenal. Russia is developing new nuclear weapons. North Korea is accelerating its long-range nuclear strike programs and India's and Pakistan's nuclear arsenals continue to grow. Depending on what occurs in Ukraine and Iran, more nations may acquire nuclear weapons options of their own.

Parallel to these developments, the military salience of space has increased dramatically. Not just the United States and Russia, but China, Europe, Japan, the Koreas, Israel, and India have begun to develop military space capabilities. Just as many states are planning missions to the Moon. Finally, the development of dual-use commercial space systems that can be used for military purposes is expanding significantly.

The space wargame described below focuses on the collision of these two trends by simulating North Korea's use of nuclear weapons in space. Initially, some of the game's participants thought this was quite unlikely; mid-way through the game, though, they warmed to the idea. Some suggested that China might exercise the option.

They were prescient. Twelve weeks after NPEC completed its simulation, *The South China Morning Post* reported the People's

Liberation Army's Northwest Institute of Nuclear Technology had simulated a nuclear weapons strike designed to knockout dual-use satellite constellations, such as Starlink. The stated aim of the computer-simulated attack was to prevent Taiwan from exploiting such militarily useful commercial systems. The article noted that the Limited Test Ban Treaty prohibits the detonation of nuclear weapons in space and the atmosphere. It failed, however, to point out that neither China nor North Korea is parties to the treaty.

The game scenario, in short, is plausible. By the end of the decade (when this simulation's war takes place), tens of thousands of small, commercial, networked satellite systems will be flying in low-Earth orbit. These satellites will complement the U.S. Defense Department's own space architecture, which includes national security systems flying in low, medium, and geosynchronous orbits. Hostile states will want to threaten them.

What's the worst they might do? This report gives an extreme answer. Some might ask why consider it. The answer is simple: U.S. national policies frequently focus on dire hypotheticals — e.g., massive nuclear wars, global warming catastrophes, pandemics, etc. — to shape plans not only to hedge against the worst, but to deal with lesser, included threats. U.S. space policy arguably doesn't yet have such an organizing worry. The scenario this report focuses on might help.

The war described in the report begins in the spring of 2029. The DPRK tests an ICBM that inadvertently travels further than intended, triggering U.S. missile defenses. No interception is made, but the U.S. demands North Korea show a good-faith effort to avoid further provocations by garrisoning its mobile missile force and by beginning U.S. reconnaissance flights near North Korea. Washington subsequently increases pressure on North Korea to accede to these demands by asking the United Nations to approve a selective blockade of North Korea and placing U.S. strategic forces on alert.

North Korea refuses America's demands, begins a general mobilization, and warns Washington that if the United States fails to end its nuclear alert and refuses to schedule the removal of its troops from the South, war will ensue. Tensions continue to build. Then early in June, North Korea launches a satellite into orbit and warns of a possible nuclear explosion in space unless the U.S. and ROK stand down. Washington contacts Beijing in hopes of getting China to pressure the North Koreans to relent. Chinese officials counsel Washington to negotiate with Pyongyang directly, noting that North Korea has not yet violated any treaty. The United States goes to the United Nations Security Council with a sanctions resolution against Pyongyang's space provocation. Russia and China block its approval.

Throughout this crisis, U.S. officials try to determine if North Korea's satellite is actually carrying a nuclear payload but are unable to do so. In mid-June, North Korea launches another payload into space, this time over the North Pacific. Well before entering full orbit (which, if it did, would constitute a clear violation of the Outer Space Treaty), the payload detonates, releasing 10-20 kilotons of nuclear energy into low-Earth orbit. All satellites in line of sight of the explosion are immediately disabled. U.S. space experts predict that the rest of the world's satellites in lower low-Earth orbit will be disabled in a matter of days to several weeks. Tactical military communications and reconnaissance in Korea are hampered by North Korea's low-Earth orbit attack. Shortly after the detonation, North Korea invades South Korea.

Key Takeaways

At each move, the game focused on what the United States and its closest space-faring allies would do to deal with each of these crises. This produced four takeaways:

1. ***The popular notions that space warfare will stay in space and that international limits can prevent hostile actions from occurring there are both strong and wrong***. Diplomats hope that with enough rules of the road, norms, and diplomatic signaling, the worst in space — military combat — can be avoided. Yet, the strong belief that diplomatic limits might prevent hostile military actions in space is betrayed by deep ambiguities in the space restrictions we have. In this regard, the U.S. team insisted North Korea's nuclear detonation violated the Outer Space Treaty. China, however, disagreed and the U.S. Department of Defense legal experts agreed with China: Unless it can be proven that a nuclear device detonated while it was clearly in orbit or "on station," there may be no foul. Indeed, under the Outer Space Treaty, both the Chinese and Pentagon lawyers insisted that a state can legally inject a nuclear weapon into space with a missile and detonate it so long as the warhead does not fly at least one, complete Earth orbit. Of course, exploding a nuclear weapon in the atmosphere would violate the Limited Test Ban Treaty but it is legally unclear how far out the atmosphere extends. More important, North Korea never signed the treaty. Also, there is nearly no easy way to verify if an orbiting spacecraft is carrying a nuclear warhead. This makes enforcing the Outer Space Treaty's ban on "stationing" or detonating nuclear weapons in space all but impossible until the treaty itself is violated with a detonation. This suggests the diplomatic utility of clarifying what disagreements we are likely to have with hostile states on these matters even if agreement on preventing a nuclear explosion in space is not immediately possible. For decades, the United States and its allies have sought to establish clear rules whose violation has consequences. Although desirable, this is still all too aspirational. NPEC's previous China space game struggled with the lack of clear rules to limit dangerous space proximity operations as well and concluded that only rules that could be self-enforced would likely be of any utility. Unfortunately, nothing in this Korean game suggested otherwise. As for the plausible hawkish hope that with

enough investments in military space capabilities, nuclear attacks in space can be deterred or, at least, be prevented from leading to conflict on Earth, the game provided no clear proof. What the game did confirm, however, is the military imperative to develop space and other command, control, communications, surveillance, and intelligence systems that can survive such attacks.

2. ***Developing satellite hardening and constellation reconstitution options to respond to a nuclear detonation in space are obvious hedges; far less obvious is what should be done to secure such options.*** All of the game's players agreed that after a high-altitude nuclear explosion, there would likely be a race to reconstitute one's satellite constellations. There also was significant agreement about what should be reconstituted — satellites and launchers — and how — by stockpiling certain materials, satellites, launchers, and other items, as well as ramping up manufacturing and mobilization base, etc. There was far less agreement or consideration, however, about *when* to reconstitute — early, when satellite lifetimes would be limited, or later, after the radiation levels in the Van Allen Belts had declined, allowing newly inserted satellites to survive longer. Nor was there agreement about where to focus the reconstitution efforts — in upper low-Earth orbit, medium-Earth orbit, in geosynchronous-Earth orbit, or on alternative, non-space-based, terrestrial systems. It also was unclear who might win in such a reconstitution race – China or the United States – and why. Some believed that the United States and its allies had a launch and satellite infrastructure and technology lead as well as a larger mobilization base than China. Others believed China would steal a march on any reconstitution race given its much quicker acquisition times. The game also stumbled across another reconstitution problem. Most players supposed that the Russian Soyuz capsule would be available on the U.S. space station in 2029. It might not. Developing U.S. or allied escape capsules would be desirable for both government and commercial space stations and operations on the moon.

Finally, there was disagreement about how much hardening should be required. Some said it was pointless to demand this of commercial space firms, that if the U.S. government did, these firms would simply go overseas. Others said hardening should be proportionate to the risk being run (is the satellite in low-Earth orbit, near debris, able to cope with strong solar storms, radiation, etc.). Others insisted that if the commercial satellite provided government services, their government contracts could be conditioned upon meeting certain hardening requirements. Yet others said the government should help pay for such hardening. Some took the position that low-Earth orbit satellites ultimately were not that important to U.S. and allied security and prosperity. Others disagreed. What is clear is the need for our government to resolve these disagreements before a crisis approaching what the game played ever unfolds.

3. ***To hedge against possible nuclear attacks against low-Earth orbiting systems, the United States and its space-faring allies should develop alternatives both in other space orbits and on Earth.*** Devising space systems that can operate in several different orbits without major modifications or dramatically increased costs would be extremely useful to cope with the threats posed in the game. Such systems could afford the United States and its allies much greater levels of space system resiliency against both nuclear and nonnuclear threats. Such systems would also allow much greater operational flexibility in determining what an "optimal" constellation might be. Finally, it would make any reconstitution efforts much quicker and, possibly cheaper. Meanwhile, on Earth, it would be useful to fortify land and sea cable communication systems, ground-based navigational aids, and the development of alternative imagery systems mated to high-altitude balloons, long-endurance drones, or other non-space platforms. These land and air-based systems could be useful as stop gaps if low-Earth orbit space systems are disabled. In developing these alternative systems, it also would be most useful to develop ways to defend them both passively and actively.

4. ***Two large unknowns worth assessing are what can and cannot be done to verify the presence of nuclear weapons in space and what a military proportionate action might be to a nuclear detonation in space.*** There was considerable debate about America's ability to verify a nuclear payload in space. Some thought it would be possible by 2029; others doubted this. Several questioned if without a large fleet of such inspection systems constantly in orbit, one could get close enough soon enough to make any difference if a nuclear threat arose. Whatever the truth may be, it is critical to determine what's possible so that reasonable expectations can be set before a crisis occurs. It is always tempting to spend significant amounts of time and money to try to secure a technical fix for such a challenging detection mission. It may more sense, however, to design technically and militarily around the probability that such a fix may long prove elusive. Yet another mistaken assumption the game revealed was that agreed, proportionate military counterstrike options to a nuclear detonation in space would be on tap. They were not. Again, it is unclear what, if any, these options might be.

Wargame Format

The wargame consisted of three moves. The first began with a mishap during a North Korean test of multiple, independently targetable warhead-armed intercontinental ballistic missiles — this sparked an international crisis. In the second move, the crisis escalated, North Korea moved its ground forces near the border with South Korea and set off a nuclear weapon in low-Earth orbit. The game's last, third move was a "hot wash" in which the group discussed the simulation and the players' key findings

Wargame participants were organized into four teams representing the United States, South Korea, China, and an expanded Five Eyes team (i.e., the United Kingdom, Canada, Australia, New Zealand, Japan, Germany, and France). The control team oversaw communications, managed the scenario, and represented North Korea, Russia, and several other regional actors. Teams responded to the crisis; communicated with other teams to gather information; negotiated; and created a response strategy and contingency plans.

Move One – The Crisis

Move one begins in 2029. The United States and its allies have employed large networks of proliferated military and commercial satellite systems flying in low-Earth orbit. A good portion of America's, Japan's, and South Korea's most important military and civilian space missions are no longer achieved by simply using several large, geosynchronous and semi-geosynchronous spacecraft. Instead, by 2029, the U.S. Department of Defense has fielded their own fleet of military satellites and contracted with private firms that have their own large dual-use satellite constellations flying in low-Earth orbit. These low-Earth systems complement the large spy satellites the United States still operates. The low-Earth orbit military constellations' key missions include missile tracking and secure military communications. In addition, tens of thousands of low-Earth orbit, dual-use, commercial, networked satellite systems afford unclassified images and secure communications not only to the U.S. military and intelligence agencies and America's closest security allies but to customers lacking U.S. security clearances.

Early in 2029, South Korea's and America's recently elected presidents, announce their intent to meet to discuss several issues of common interest, including how best to deal with the continuing military threat Pyongyang poses. To test their resolve, North Korea launches a multiple independently targetable re-entry vehicle-armed intercontinental ballistic missile toward North America. The missile's flight is supposed to terminate before it might trigger America's Alaska-based missile defenses. Unfortunately, something goes wrong. The missile flies further than planned, triggers the

Alaskan system, but splashes down before it can be intercepted. This simulated "attack" alarms the American public.

The President demands that Pyongyang show good faith to prevent any further "surprises" by parking all of its mobile missiles in three agreed, visible locations within North Korea. In addition, Washington announces it will ask the United Nations Security Council to "prevent further military provocations" by authorizing a selective air blockade of North Korea. China and Russia block consideration of this resolution and before any further diplomatic action is taken, North Korea mobilizes its military. In response, the United States places its forces on alert, consults Seoul and Tokyo, and orders a series of air reconnaissance flights close to North Korean air space.

This escalates the crisis. North Korea warns that unless Washington 1. stops these flights, 2. stands down its military, and 3. agrees to a scheduled withdrawal of U.S. troops from the Peninsula, war will ensue. On the same day, Pyongyang launches a road mobile solid-fueled rocket but instead of aiming it against a target on Earth, the missile places an object in low-Earth orbit. North Korea, then, announces that Washington and Seoul should be very worried that a nuclear explosive may be set off in space at any time.

In hotline consultations with China, Beijing reminds Washington that North Korea has not yet clearly broken any rules as there is no reason to believe that it has placed an object carrying nuclear weapons in space. The PRC calls for calm so it might host talks to resolve the "larger" crisis. It offers to arbitrate and warns the United States not to attack North Korea's "satellite" with anti-satellite weapons and that it should let diplomacy run its course.

Given this communication, the United States decides against, taking preemptive military action. Washington decides instead to keep engaging China, South Korea, and America's allies to harmonize their actions if the situation intensified and asks Beijing to talk directly with Pyongyang. U.S. officials stress that they would hold

North Korea's leadership accountable should the crisis further escalate.

Washington gives its diplomatic strategy several weeks to succeed and concurrently prepares for its possible failure. The United States continues its military reconnaissance of North Korea and explores quick-fix technical workarounds to support military operations that rely on space assets that might be lost if North Korea sets off a nuclear weapon in orbit. The United States halts any new U.S. launches of satellites into low-Earth orbit and reduces the crew on its manned space station to two, which is the minimum required for operation.

South Korea considers North Korea's actions an act of war that can only be de-escalated by showing Pyongyang an appropriate level of allied military resolve. After meeting with South Korea's president, the U.S. President places U.S. nuclear strategic forces on Defcon 3 alert, orders families and dependents to evacuate from the Korean Peninsula, and calls for the U.S. embassy to reduce its staff. South Korea expresses concern that North Korea's detonation of a nuclear weapon in space could cause substantial harm to South Korean communications, surveillance, early warning, and precision targeting systems.

Meanwhile, China perceives the crisis as one of America's making and characterizes the U.S. response as a deliberate escalation. China says it will work to de-escalate the crisis and discourage North Korean nuclear use but that it will not support any measures that impinge on North Korea's sovereignty, including imposing United Nations sanctions or blockades. China also issues a public statement defending North Korea's exercise of "its right to test defensive weapons" and calling for a meeting of all parties to find a peaceful resolution to the current crisis.

In this regard, China reminds the United States of its perpetual treaty of friendship with North Korea and maintains that North Korea's sovereignty is a fundamental Chinese interest. Some believe China sees the crisis as an opportunity to drive a wedge between America and its partners. China, meanwhile, offers to convey to Pyongyang Washington's desire to deescalate the crisis. Toward that end, Beijing advises the United States to de-alert its nuclear forces and agree to three-way talks with China and North Korea. At the same time, China asks North Korea to give China advance notice of any nuclear detonation to ensure the safety of astronauts on the Chinese space station and protect vital space infrastructure.

Both the South Korea and expanded Five Eyes nations reject China's characterization of the United States as the aggressor and condemn North Korea's actions. South Korea urges China to exert its influence over North Korea to maintain peace and warns that Seoul will increase defensive measures as long as North Korea continues to threaten the South Korean people. The South Korean team advises Beijing that South Korea's actions are of a defensive nature and not meant to undermine China's efforts for peace. South Korea and the expanded Five Eyes nations also publicly remind all nations of the collective threat a North Korean detonation of a nuclear weapon in space poses to Chinese, Russian, and Western commercial satellites and civil space stations operating in low-Earth orbit.

The expanded Five Eyes nations evacuate personnel from their manned space stations and the moon to prevent them from being stranded should the stations or the command and control to them be damaged. The expanded Five Eyes nations also begin migrating communications, navigation, and imagery activities in low-Earth orbit to satellites operating in other orbits, where feasible.

Meanwhile, China applauds South Korea's recent space accomplishments, cautioning that China would not want to see those assets affected by any escalatory actions. Beijing implores Seoul to refrain from military actions that North Korea might perceive as threatening.

The U.S. team directs American commercial satellite firms to do what they can to inspect the orbiting North Korean satellite and prepares for possible non-kinetic actions against North Korea's satellite (i.e., with high-power microwave and laser targeting and jamming). Meanwhile, the U.S. team asks China to support a United Nations resolution calling on North Korea to abide by past United Nations resolutions and end its nuclear provocations.

In response, China issues a firm warning to North Korea that it is in nobody's best interest to escalate the crisis. China reminds North Korea of the interconnectedness of their two economies and how any actions that cause harm to the Chinese economy, would, in turn, be detrimental to North Korea.

Move Two – The Detonation and Aftermath

At the start of move two, the crisis is in its 12th week. On June 10 and 11, the DPRK prepares four road-mobile missiles for launch and raises its force's alert status. On June 12, North Korean ground forces move toward the Demilitarized Zone. U.S. intelligence assesses the potential for an invasion to be high. On June 13, North Korea activates multiple mobile missile batteries, transiting to various points at night.

Meanwhile, the results of the U.S. satellite inspections begun in move one reveal gamma ray emissions from North Korea's satellite's payload, suggesting the presence of some plutonium. However, the inspection is unable to determine whether the payload is a weapon. South Korea asks the United States to conduct a preemptive space strike to stop North Korea from detonating the satellite. However, Seoul because such an action would be considered offensive and because a kinetic attack would cause damaging debris, Seoul urges Washington to use non-kinetic anti-satellite weaponry (e.g., jamming or cyber).

Meanwhile, America's space fairing allies ask for U.S. assistance to defend their assets in low-Earth orbit. To prepare for a possible nuclear detonation, the U.S. team urges contractors to determine key choke points and to prepare to reconstitute their constellations.

Meanwhile, the South Korean team partially mobilizes its armed forces and places 75 percent of South Korea's forces under U.S.-South Korean Operational Control or Central Forces Command. To hedge against a possible North Korean invasion, Seoul asks Washington to deploy one to two guided missile submarines off the Spratley Islands, one or more attack submarines in Taiwan Strait, and a carrier group off Japan and the Philippines to distract China and keep it out of Taiwan.

Washington assures Seoul that the United States is taking appropriate preparatory measures to prevent a Chinese move against South Korea and Taiwan asks Seoul to prepare to move north, with U.S. forces, depending on North Korea's and China's next move. The United States assures Seoul that major, U.S. strategic space systems in medium and geosynchronous orbits, such as Defense Support Program Satellites, MILSTAR, and GPS are hardened and are fully operational but communications and imagery systems in low-Earth orbit will degrade over time due to the increased radiation environment.

The U.S. and its allies strongly condemn North Korea's actions that have broken the Armistice and that the firing of any rocket by North Korea is a breach of existing United Nations resolutions and Article 9 of the Outer Space Treaty, which prohibits "potentially harmful interference" with legitimate activities in space. They ask the United Nations to call out North Korea for violating the Outer Space Treaty. As a precondition to engaging in any Six-party talks, Washington insists that North Korea first commit not to explode a nuclear device in space. The Five Eyes expand their surveillance to detect any move China might make to take advantage of the space crisis to act against Taiwan. Privately, Beijing warns North Korea that a nuclear detonation would harm both China and North Korea. Pyongyang replies that only the withdrawal of all U.S. forces from the Korean

Peninsula and America's commitment to the preservation of the North Korean regime can de-escalate the crisis.

On June 14, North Korea test ripple-fires four missiles. None go into orbit but one, carrying a nuclear device, detonates over the Northern Pacific well above the atmosphere. The device appears to have a nominal nuclear yield. On the ground, North Korean artillery forces bombard military targets just beyond the Demilitarized Zone. North Korean light infantry forces penetrate the Demilitarized Zone and use special forces to attack multiple air bases in South Korea.

Most U.S. and allied commercial satellites within line of sight of the nuclear explosion either immediately fail or begin to fail, including communication, entertainment, internet, weather, and Earth observation systems. The crews remaining on Chinese, Russian, and U.S. space stations are exposed to high levels of radiation. Initial assessments of debris creation are high. There is no electromagnetic pulse effect on any territory of any nation.

The United States uses ground-based, high-powered microwave and laser systems to disable North Korea's orbiting satellite, creating no debris, and activates its Korean command, executing all standing operational plans. Washington holds the North Korean regime accountable for launching a war against South Korea and proposes a United Nations Security Council resolution condemning, in unequivocal terms, North Korea's use of nuclear weapons in violation of the Outer Space Treaty. The Americans implore China to restore peace on the Korean Peninsula and avoid escalatory activities outside the Peninsula. Washington says it has no plans to use nuclear weapons, but warns that America and its allies will take all steps necessary to ensure North Korea never uses nuclear weapons again and to prevent North Korean forces from successfully invading the South.

South Korea's highest priority is to defend itself against the North Korean invasion. It initiates the operational plan and moves troops north, asking the United States to confirm it will provide air and naval capabilities to support the conflict. The United States assures South Korea and its other allies that it will provide the logistic support necessary for Korea to win the conflict. The U.S. also deploys naval forces to deter or defeat any attempts by China to take Taiwan.

South Korea objects to Washington's announcement that the United States would not use nuclear weapons. For Seoul, "communists only understand strength." At the same time, South Korea investigates how it might continue to conduct military communications, surveillance, and navigation by relying on ground and air-based systems. They also inquire if there are functional American space systems that South Korea could rely on outside of low-Earth orbit.

The United States, the Five Eyes, Japan, Germany, and France form a standing group to assess the military and civilian operational implications of the satellite losses. The U.S. team investigates what can be done with commercial space companies to surge satellite production and accelerate launching replacements. The U.S. team also conducts an inventory of what responsive launch operations capabilities it might have that allow the launching of satellites from planes based at airports. It also increases security around fixed launch facilities at the Kennedy Space Center and Vandenberg. The U.S. team is also concerned about possible attacks against U.S. space-related production lines and non-space communication systems (i.e., underseas cables).

Meanwhile, Five Eyes, Japan, Germany, and France use backup systems in medium-Earth and higher obits to restore as much communication as possible. They also increase the monitoring and protection of undersea cables and the use of high-altitude airborne systems (e.g., high-flying drones, balloons, etc.) and terrestrial system to help observe the war on the Korean peninsula and monitor Chinese military movements. China does the same, announces its support for North Korea, and warns South Korea and its allies not to cross the 38th parallel. The emergency of war makes forces deferral of any immediate clarification of what, when, and how the United States and its allies will reconstitute their commercial, civil, and military low-Earth orbital space systems.

Commercial Satellite Use Catalyzes Nuclear-Armed States to Combat: A Wargame After-action Report

June 2023

Ukraine's successful military use of Starlink in its defense against Russia has become a poster child for nations' increased military reliance on commercial space systems. Yet, military use of commercial space services could create as many problems as it solves. *The New Yorker* recently investigated Ukraine's use of the Starlink program. This coverage raised a disturbing question. Might private space barons and state-controlled commercial space firms — e.g., Elon Musk, Jeff Bezos, One Web, Maxar, and Guo Wang/StarNet — switch off these critical systems, jeopardizing the security of major and lesser states?

This question is but one of many. Could an attack against a commercial space system constitute an act of war? Might it, in turn, escalate to all-out nuclear war (a 21st-century space Lusitania)? What of space being a truly separate warfighting domain? Some argue that space isn't a separate warfighting domain but merely an arena for conducting support operations for ground, air, and sea warfare. Supporters of this view insist that whatever military actions are taken in space stay in space. Are they right? Or is space a domain in which space combat action itself can further catalyze ground, sea, or air engagements, as well as further space combat?

Should governments assume responsibility for defending commercial space systems? When, why, and how should they do so?

Should governments remunerate commercial space system companies if an adversary damages their system? Or should wartime damage of space-based systems simply be considered a part of doing business — handled as other risks are with light national regulations and private insurance? How, if at all, should governments facilitate such regulation? Is there a clear line between peacetime and war operations in space? What if a non-state actor uses commercial space systems to conduct attacks? Who, if anyone, should be held accountable?

If the answers to these questions are less than crystal, should Washington lay down minimum hardening or reconstitution requirements for all commercial satellite systems, for companies contracting with the U.S. government? What role, if any, should international governance play in reducing the risks of military conflict and escalation? Also, what, if anything, should the U.S. and allied governments do to protect satellites (both commercial and military) or to enforce any rules that states might want? Are additional passive and active defenses needed; what about space bodyguard systems that could gently push potentially hostile robot spacecraft a safe distance away from critical satellites?

So far, the United States and other space-faring nations have either mostly deferred or taken a hands-off approach to dealing with these questions. Instead, Washington has treated the commercial development of space as it has the internet — as a special area in which protecting First Amendment rights and free-trade entrepreneurialism trumps instituting new forms of regulation. With space, states are roughly where they were with sea power in the 1600s and air power before World War One. In these periods, dramatic acts of air war and naval piracy were about to ensue but instituting national or international regulation hardly seemed urgent… until they were.

All of this raises a fundamental question: Just how sustainable is it for us to continue our current relaxed approach to the military exploitation of commercial space?

Game Play

To find out, NPEC designed and conducted a wargame this summer tailored to this purpose. The game's play begins in 2027. India contracts with Maxar, a U.S. commercial imagery satellite company to buy a controlling 51 percent interest in this system's three-satellite, sun-synchronous system (a constellation that is eventually to be filled out to six-satellites). Washington backs the sale and not only buys New Delhi free mobile ground stations, but blesses India controlling a 51 percent share, and places U.S. military payloads on the system to demonstrate the significance of this U.S.-Indian space collaboration.

Meanwhile, China develops a similar system and sells its imagery and communications links to a Pakistani government-blessed private entity. The entity is run by retired Pakistani General T. Khan, who has close ties to Pakistani terrorist groups. Without asking for explicit Pakistani permission, the General gives one such group access to the Chinese satellite system's "peaceful" commercial imagery and communications links.

The terrorist group wants Islamabad to take a stronger stand against India's "occupation" of Kashmir. Towards this end (i.e., to force the hand of Pakistan and draw it into a major war), the terrorist group mates long-range drones it previously had acquired with the Chinese satellite service imagery and the service's secure communications links to strike India's strategic nuclear air base at Ambala twice. The terrorists' drones destroy several Indian nuclear-capable Rafale fighters and NETRA strategic nuclear forces command planes and kill dozens of Indian airmen.

U.S. intelligence and other open sources confirm that China's satellite system supported the strike. A flurry of diplomatic and international commercial legal initiatives ensues. None, however, are ultimately acted upon. Under mounting domestic pressure to act, India uses a newly constructed ground-based space-tracking laser to dazzle the Chinese-Pakistani satellite system the terrorists used. This dazzling unintentionally damages the Chinese satellite's optics.

China, in the game's second move, retaliates by using its own ground-based laser systems to damage the optics of one of the U.S.-India Maxar satellites. Shortly after, a Chinese rendezvous satellite closes in on a second U.S.-India Maxar satellite as it comes within range of China's laser system. This second satellite, including its American payload, goes dead. It is unclear what caused the satellite to stop functioning.

Meanwhile, India informs Washington that another Chinese rendezvous satellite is closing in on the third and last U.S.-India Maxar satellite. Shortly thereafter, this Maxar satellite also goes dead. As with the previous Chinese attack, the United States has no bodyguard satellites to deflect a possible Chinese rendezvous satellite assault. At this point, U.S. intelligence briefs the President, who authorizes a covert U.S. cyber-attack against the offending Chinese rendezvous satellite, which disables it. The Chinese immediately blame Washington for killing their satellite.

Meanwhile, India launches a cruise missile attack against suspected terrorist sites in Pakistan. As the Pakistani government ponders what retaliatory military action it will take on the ground against India, the U.S. Space Command readies itself for a Chinese space counterattack.

Findings

The game's final hot-wash discussion session supported four key findings:

1. ***Space combat can catalyze combat both on Earth and in space.*** Many experts like to believe that whatever happens in space stays in space. This game strongly suggested otherwise. India and Pakistan's exploitation of commercial space satellite systems not only encouraged the United States and China to attack each other's commercial satellites but intensified land warfare between India and Pakistan. What makes this finding worrisome is the continued lack of clarity as to what an act of war might be regarding commercial satellite systems that get "damaged" or are exploited for military purposes. Nor does it help that such military space operations can be conducted quickly and send complex, ambiguous signals to military and civilian space operators. In the game, no fewer than four nuclear-armed states struggled to determine who was doing what, and even what was happening, as India's nuclear strategic forces were seriously degraded and a U.S. military space payload was destroyed. All of this could be quite escalatory. At a minimum, it recommends engaging all space-faring nations in further talks to clarify what acts of war in space might be and determine how their detection might best be enhanced and verified. The latter would likely entail some combination of private national and public international efforts. Ideally, one might create a dedicated international body to verify illicit space activities that could credibly assign attribution. This international body's surveillance requirements, in turn, could be used to help justify additional national funding of private space support contracts to increase the quality and availability of space situational

awareness information. It is unclear what, if any, thinking is being done within the U.S. government or elsewhere to determine what the optimal mix of private and international space surveillance and verification efforts should be. Beyond this, the United States and its space-faring partners need to consider what, if any, new kinds of space capabilities, such as bodyguard satellites and active satellite defenses, might be needed to enforce desirable offensive space operations red lines.

2. ***As commercial satellite systems spread, rogue states and terrorists will try to exploit them, increasing the likelihood that major, nuclear-armed states could be dragged into wars.*** In the game, a terrorist group based in Pakistan uses commercial satellite systems to target strategic Indian nuclear assets with the aim of forcing the Pakistani government to side with them and wage a major war over Kashmir. What made this gambit relatively easy was the largely unregulated provision of "private" commercial satellite services and links to a wide variety of states, firms, and non-state entities. At a minimum, this suggests the United States and other like-minded governments should encourage private firms operating under national jurisdiction to assume greater responsibility for their possible misuse. In specific, space-faring nations should explore creating commercial rules analogous to "know-your-customer" rules used in the banking industry. Governments should apply these new rules to private space service providers, holding them responsible for the harm their customers inflict using their services in wars or through acts of terrorism. The challenge here will be to prevent smaller nations from offering licenses under little or no such regulation. In these cases, insurance premiums from reputable insurers should be set much higher than for properly regulated licensing or not

be made available at all. Again, it is unclear to what extent governments and private firms are yet thinking through these matters.

3. ***With the increased use of commercial satellite systems, America's credibility in mediating conflicts between nuclear-armed combatants will be questioned in new, demanding ways.*** Historically, Washington has acted as an honest broker in numerous Pakistani-India military crises. In this space scenario, the United States, however, deferred to India so much that Islamabad and Washington mistakenly concluded there was little point in engaging with one another and that India was free to take considerable action on its own. This resulted in Pakistan reluctantly conceding to heavy-handed Chinese bullying for Beijing's pre-clearance of any Pakistani diplomatic crisis messaging. If Washington had done more to engage not just India, but Pakistan, on security issues prior to the conflict, this might have been avoided. Once conflict began, though, both India and Pakistan sought Washington's information on what was occurring in space and on the ground. Unfortunately, neither Pakistan nor India knew how much Washington knew and Washington's lack of transparency created distrust. Because space warfare and its diplomatic management demand more space situational awareness than will ever be available, building trust is vital. One way to increase confidence and transparency is to encourage the private sector to provide more access to the space situational awareness information they might have (something the American team facilitated in the game). Governments, including the United States, might facilitate this by paying private firms to share what they know with international or multilateral space organizations that, in turn, would make it generally available both in peacetime and during crises. Another way to increase trust,

that the game players discussed, would be to create new multilateral space security working groups starting with states in war zones (including those that have nuclear weapons on their soil) that Washington has working relations with (e.g., three-way talks among the United States, Pakistan, and India; with Turkey and Greece; among Middle Eastern states and Israel, etc.).

4. ***Leaving space activities as unmanaged as the internet is a prescription for military mischief.*** One of the major game discoveries is that satellite systems' increasing duality makes it difficult to attribute the cause of destructive and disruptive space incidents or to determine appropriate responses. In the game, the United States puts a major U.S. military payload on a commercial spacecraft that India controls. China attacks it; the United States conducts a space counterattack, escalating the conflict. This play raised several questions. What military missions should only be conducted on government-owned dedicated military spacecraft? If the United States needs or wants to place military payloads on foreign-owned spacecraft, should it only do so if the United States has a military security agreement with the state from which the spacecraft is launched? Should governments or international organizations require a minimum amount of survivability features for all commercial satellites (especially those that are dual-use)? Should states condition any government protection or indemnification of private space assets against foreign military assaults? What should these conditions be — hardening, adequate insurance, being able to quickly reconstitute the targeted space system, supporting the deployment of bodyguard satellite systems to deflect possible hostile rendezvous satellite assaults, etc.? All of these questions were raised before or during the game; none were answered.

Wargame Drivers

There were several drivers pushing the game's play. The first was America's desire to befriend India further through space cooperation on the sale and management of a commercial sun synchronous Maxar-satellite service. In addition to buying New Delhi free mobile ground stations, Washington blesses India owning a controlling 51 percent of the project, and places a U.S. military payload on the system as a further demonstration of U.S.-Indian trust. Throughout the game, the United States wanted to be neutral but found itself tilting toward India.

At least as important a game driver was the agenda of the Pakistani players. The Pakistani terrorist group wanted to get Islamabad to back the terrorist organization's demand to threaten war over Kashmir and to exploit the Pakistani government's need to placate this group. Towards this end, Islamabad gives the group access to commercial Chinese satellite imagery and the secure communications services Pakistan has purchased and put under the supervision of a retired Pakistani general and terrorist sympathizer, General T. Khan. Even though Pakistan no longer has the pull it once had on Washington because of the Afghanistan withdrawal, Islamabad wants to maintain good relations with the United States as well as China. Pakistan does not seek war with India.

Yet another driver of the game's play was India's desire to rectify the threat Pakistani terrorist attacks posed to India's strategic nuclear forces. India, like Pakistan, also had a clear desire not to be drawn into any larger disputes that the United States and China might have, particularly regarding attacks on Chinese and American satellites.

Finally, the last key game driver was China's desire to control Pakistan's actions against India and its diplomacy more generally. Throughout the game, China (played by control) uses the crisis as a zero-sum contest to increase Beijing's control of space at the expense of U.S. control and influence.

Move One

Late in 2027, India seals a business contract with Maxar that allows India to gain access to this company's latest sun-synchronous system, which consists of three satellites (eventually to become six). This entails Maxar becoming a 49 percent investor in a new Indian company. As with the previous Pentagon purchase of Starlink terminals for Ukraine, the U.S. Defense Department contracts with Maxar to buy mobile Maxar terminals for India. The Maxar satellites involved include space situational awareness capabilities and a payload important to the United States. The deal, which has dual-use military implications, has the blessing of the U.S. government.

A Chinese company closely linked to the CCP, meanwhile, develops a similar system and sells direct uplink and downlink connection, as well as imagery analysis services, to a Pakistani "private entity." The Chinese constellation is capable of both remote-sensing and secure satellite communication data links. These capabilities would give the user command prompts for any navigation required to get to a given target, allowing over-the-horizon command of remote assets. The Pakistani "private entity," is owned by retired General T. Khan, a former Inter-Services Intelligence (ISI) official who allows terrorist groups based in Kashmir to access this commercial resolution service. Islamabad turns a blind eye to this arrangement.

The terrorists are eager to force the Pakistani government to commit fully to the terrorists' cause in Kashmir. One of these groups takes advantage of its access to the commercial imagery and satellite data link to use drones to target India's nuclear-capable Rafale fighters at Ambala Air Force Base (Pakistan famously attacked this airbase in 1965 and 1971) and the NETRA Airborne Early Warning and Control planes that are temporarily forward-based there. The terrorist group launches its attack against these targets and successfully destroys two Rafales and one NETRA plane, killing five Indian airmen servicing the NETRA aircraft.

India's military scrambles to determine the source of the attack, reaching out to Washington and U.S. intelligence services. The CIA concludes that Pakistani-contracted Chinese satellite imagery services supported the attack. Given this information, the Indian prime minister immediately goes on television identifying Pakistan as the perpetrator. India demands a Pakistani apology, admission of guilt, and a crackdown on terrorist groups based in Pakistan. India also accuses the Chinese of providing satellite imagery to support the operation.

Islamabad denies it had anything to do with the attack. Beijing states that it bore even less responsibility. The satellite service, Beijing noted, was the result of a private commercial contract between Pakistan and a private Chinese satellite company. India, which sits on the UN Security Council, asks for an emergency meeting.

Shortly thereafter, the terrorists launch a second attack against the Ambala Air Force Base. This attack takes out another three NETRA command aircraft and results in 24 fatalities and at least 15 other major casualties. India again accuses Pakistan and China of allowing the use of Chinese satellite services to pull off this attack.

Facing major internal protests, India decides to show its displeasure unilaterally: it uses its recently built laser tracking system to dazzle one of the satellites in the Pakistani-Chinese constellation. It holds in reserve its direct ascent capability, cyber, and local jamming capabilities. Unintentionally, it damages the satellite's optics. This is the first dazzling to result in permanent optic damage.

China and Pakistan condemn India. China demands India pay for China's damaged satellite and that India dismantle its commercial laser tracking system. China threatens to take proportionate action if its demands are not met.

India rejects the demands, calls for arbitration at the UN Security Council, and pressures the United States to give a detailed brief on its intelligence regarding the attack. The U.S. intelligence community is wary to do so. It has human intelligence showing that Pakistan was aware of and did not prevent terrorists from using China's satellite services to support both attacks but insists that sharing this evidence could jeopardize the life of a critical intelligence source. A debate within the U.S. policy and intelligence communities as to what to do ensues.

The U.S. team gathers additional intelligence to determine specifics about how the attack was carried out and attempts to confirm the link between China and Pakistan and the terrorist organization. Indian intelligence confirms the attack was carried out by long-range, over-the-horizon drones capable of carrying up to 75 kilograms of explosives. India claims the drones were launched by Pakistani-backed terrorists using imagery from the Pakistani-Chinese satellite system to ascertain the location of large moveable planes at Ambala Air Force Base and its communication links to queue up the attack. China continues to insist the imagery was provided by a private company.

Pakistan publicly demands that India release its intelligence confirming Pakistan is responsible for the attacks. Privately, Pakistan asks China how it will respond to the destruction of its satellite and whether the launch of the co-orbital satellites is part of this response. Pakistan also asks Beijing to back a statement condemning India for jeopardizing space security. China responds that "it always supports its close ally, Pakistan. Regarding the satellite launch, China upholds its responsibilities under the Outer Space Treaty to ensure no entities use their space assets for non-peaceful purposes and remains committed to a rules-based order in outer space including the inherent right to self-defense."

Pakistan separately asks the United States whether it intends to be an honest broker in resolving the crisis, and if so, why Washington is sharing intelligence with India. The United States affirms its commitment to ensuring the peace and stability of both South Asia and outer space and states that it is working to understand the details of both the attacks on the Ambala Air Force Base and incidents in outer space. Washington also asks Pakistan to share any intelligence it has about the crisis.

Meanwhile, India asks the United States for intelligence updates regarding the involvement of Pakistani proxy groups in the attacks on the Ambala Air Force Base, indications of additional drone sorties, and whether there were satellite downlinks to the drones. India issues a strong statement to Pakistan condemning the targeting of its nuclear capabilities and implores the Pakistani government to address the proxy groups operating within its borders. India believes the attacks on its nuclear command and control undermine strategic deterrence and threatens to take action if Pakistan does not hand over the terrorists.

The U.S. Space Force then detects the launch of three co-orbital Chinese "space debris removal rendezvous satellite systems" and determines that they are headed into orbits similar to that of the Maxar constellation, which China previously claimed contained space debris that needed to be removed. The U.S. Representative to the United Nations issues a press release condemning the terrorist attack against Indian nuclear forces as posing a threat to regional peace and global nuclear security. In addition, the United States reminds nations of their responsibility under the Outer Space Treaty to ensure that their own entities do not use space assets for non-peaceful purposes. The United States confirms its commitment to a rules-based order in outer space including the inherent right to self-defense.

India supports the U.S. call for a rules-based order in regard to the Outer Space Treaty. As a member of the UN Security Council, India brings a resolution to condemn the attack on Indian sovereignty and calls for a resolution to constitute the attacks as a threat to international peace and security. The United States supports India's resolution. India also demands that China stop providing all space services to all Pakistani entities.

Pakistan shares India's earlier statement in which India threatened to take unilateral action against terrorists within Pakistan, with the United States, asking what Washington will do to restore peace in the region after India's aggressive attack on the Pakistani-Chinese satellites and threat to invade Pakistani territory. Pakistan then issues a public statement calling on India to halt any further attacks against civilian space assets and engage in peaceful dialog to resolve the crisis. Pakistan also asks India to share credible evidence that the terrorist attacks were supported by the Pakistani government. When it receives no response from Washington and learns of U.S. support for India's resolution at the Security Council, Pakistan shares India's threatening statement with China and asks Beijing to veto the UN Security Council resolution.

India requests that Washington issue a public statement corroborating the human intelligence indicating that Pakistan is responsible for the terrorist attacks. New Delhi also asks the United States to privately warn China to keep the three co-orbital satellites very far from the U.S.-India Maxar satellites. The United States refuses to jeopardize its human intelligence assets, continues to investigate the attack, and develops a course of action to prevent escalation. The United States assures India that it has reminded China that the Maxar satellites are a joint U.S.-India venture and hold significant economic and security interests for the United States. Washington makes clear that it views any interference with these satellites as negatively affecting U.S. interests. Meanwhile, the United States postures electronic and cyber assets to receive indications and warnings of imminent attacks and readies capabilities to suppress in-progress attacks against nuclear-capable forces, noting that the prevention of nuclear escalation is a peaceful purpose.

The United States eventually responds to Pakistan, asking what steps Pakistan is taking to address India's concerns and prevent further escalatory attacks, and offers to organize talks with India. Frustrated by the lack of evidence from both India and the United

States confirming the connection between Pakistan and the attacks, Pakistan asks the United States to emphasize the importance of commercial space security, especially for developing nations, at the United Nations.

New Delhi is increasingly aggravated by Pakistan's refusal to cooperate and the threats against its nuclear capabilities and faces pressure from the Indian public to respond to the death of 29 Indian military servicemen. India calls for Pakistan to provide intelligence that identifies the terrorists responsible for the attack, secure the satellite data Pakistan is receiving from the Chinese satellites, and provide evidence that India's dazzling resulted in irreversible damage. In addition, India demands that retired Pakistani ISI General T. Khan be turned over for questioning because of his association with terrorist groups. Pakistan does not comply.

Concerned that a Chinese counter-space action against the U.S.-India Maxar satellite constellation will compromise the U.S.-hosted payloads, the U.S. National Space Defense Center (NSDC) is ordered to posture U.S. counter-space assets to prevent a PRC counter-space action. The NSDC, working with the Combined Space Operations Center, stands up operations with Maxar commercial integration and the Space Information Sharing and Analysis Center to monitor PRC operations. However, the majority of U.S. team members do not want to escalate the situation any further.

Planet, HawkEye 360, and a U.S. cyber security corporation confirm to CNN that the drone attacks originated within Pakistan and the drones received continuous radio frequency communication from the Chinese-Pakistani satellite constellation. Further, the Pakistani constellation was commanded to take high-resolution imagery of the Ambala Air Force base directly before the terrorist attacks. The CEOs of the three U.S. corporations call this an "open and shut case that the terrorists used the Pakistani constellation for both imagery and command and control." A private U.S. attorney specializing in space law states, “Pakistan is clearly responsible under the Outer Space Treaty for how its entities behave in space, and this is clearly not a peaceful purpose… India can seek redress under the liability convention.”

Move 2

In response to the damage done to its satellite by India’s laser system, China mobilizes its own ground-based laser systems and damages the optics of one of the U.S.-India Maxar satellites. Shortly after, a Chinese rendezvous satellite closes in on a second U.S.-India Maxar satellite as it comes within range of China’s laser system. The second satellite, including its American payload, goes dead, this time permanently. It is unclear what caused the satellite to stop functioning. U.S. intelligence briefs the President.

The United States and India publicly complain. China says it did nothing, but notes that there may have been a collision as both the United States and India are irresponsible in how they direct their satellite. Meanwhile, one of China’s co-orbital satellites closes in on yet another U.S.-India Maxar satellite. Officials in New Delhi and Washington are both independently aware of this threat.

A debate breaks out in the Indian government on what counter-space steps it should take, including whether to conduct a direct ascent kinetic ASAT attack against a variety of possible Chinese targets. U.S. signals intelligence picks up a portion of this discussion. Simultaneously, a debate breaks out in Washington about what, if any, counter-space steps the United States should take against China. These include jamming, laser dazzling, and cyber-attacks. The United States still does not have a dedicated fleet of rendezvous bodyguard satellites and its unmanned space plane (the diminutive experimental X-37B) is not in orbit.

Pakistan condemns India for its destabilizing military space actions starting this escalatory path and calls for de-escalation and four-party talks between India, Pakistan, the United States, and China. New Delhi responds that India acted proportionally to maintain the integrity and resilience of Indian strategic deterrence. India also calls out Pakistan for its failure to respond to the terrorist attacks originating from Pakistani territory. Pakistan responds that it is conducting its own investigation into the origins of the attack.

Meanwhile, China privately demands that Pakistan make no communications with Washington or India without first clearing them with Beijing. If Pakistan refuses , Beijing threatens to stop supporting Pakistan at the UN or any other international forum regarding the current crisis. This demand is intercepted by U.S. intelligence. India, however, knows nothing of it.

In response to India's accusations, Pakistan releases a public statement that it does not clear with China beforehand. Pakistan, unsatisfied with the assessments made by American private companies, again requests that the American and Indian governments share intelligence, including raw data, to prove that the attack actually originated in Pakistan and again blames India for starting the escalation in space. China tells Pakistan that it will no longer support Pakistan diplomatically, given Pakistan's failure to pre-clear its statement with Beijing. This deeply alarms Pakistan, which privately reiterates its desire to coordinate closely with Beijing but explains its concern that pre-clearing every communication will make it impossible to manage such a rapidly unfolding crisis in real time. China does not budge on its demands. Because of this disagreement, China refuses to share intelligence on the potential impact on Pakistan's space capabilities that the ongoing space conflict might have.

The United States, now aware of China's demand that Pakistan pre-clear its communications, privately tells Pakistan that, unlike Beijing, Washington will respect Pakistani sovereignty in deciding with whom it wants to communicate and when. This message backfires as Pakistan is outraged at the revelation that U.S. intelligence is intercepting its secret messages.

In order to both repair the relationship with China and to communicate without American signals intelligence overhearing, the Pakistani Prime Minister offers to come to Beijing. Pakistan also gives China advance notice that Pakistani security services will bring in General Khan, the owner of the private entity accused of sharing satellite data with the terrorists, for questioning and offers to let China question him as well.

Meanwhile, India and the United States coordinate more closely. India shares its intelligence assessment that the Chinese rendezvous satellite closing in on the third U.S.-India Maxar satellite is intending to attack. Shortly thereafter, the satellite suddenly goes dead. Meanwhile, the United States shares its intent to impose economic and travel sanctions against those involved in China's counter-space actions so that the relevant Indian agencies can follow suit. New Delhi and Washington also agree to collaborate on a UN General Assembly resolution condemning China's counter-space actions.

The United States issues a notice to space operators to be aware of a potential orbital debris hazard from the lost U.S.-India Maxar satellite and warns operators that there is a risk from anti-satellite lasers when flying over China. Washington also publicly announces its assessment that China's actions were deliberate violations of the Outer Space Treaty. The United States does not take any direct action against the Chinese rendezvous satellite.

Meanwhile, the story is picked up in the media. On *Al Jazeera*, LeoLabs and HawkEye 360 provide an animation showing radar targeting of the Indian Satellite, geolocation of a large radiofrequency pulse consistent with open-source imagery of a Chinese laser facility, and the immediate change in telemetry and orientation of the satellite, followed by a rendezvous-proximity operation touching and ripping off a solar panel and U.S. sensor from the Indian satellite. A cyber analysis firm is able to show which Chinese ground station actively controlled the co-orbital onto the target. *The Guardian* reports an anonymous American intelligence official saying "it's absolutely clear that China first lased, and then tore apart the Indian satellite, taking our hosted payload with them." The Heritage Foundation releases a statement urging the Administration to show strength in the face of Chinese aggression. The Secure World Foundation and Union of Concerned Scientists urge all parties to show restraint to avoid heading toward uncontrolled space warfare and debris that could foul the space environment for generations. Elon Musk tweets "This is naked aggression against a commercial space entity!"

A spokesman for the Chinese Foreign Ministry retorts that these anonymous and non-governmental sources are providing disinformation that the U.S. government is uncomfortable validating. Beijing calls these reports a typical example of American colonial manipulation and calls on U.S. intelligence agencies to comc clean regarding any intelligence they have. The U.S. Secretary of State responds that China's behavior is a violation of the Outer Space Treaty and a betrayal of China's own proposals to create a treaty banning the placement of weapons or threat of use of force in space. India publicly supports this statement. China accuses India of tying its fate to Washington's military coattails and parroting disinformation that even the U.S. government is unwilling to validate. While the U.S. intelligence community does not publicly release its evidence, Washington confirms that the open-source analysts' findings align with the government's.

India announces that it is seeking indemnification through Article 3 of the Space Liability Convention for the loss of both of its Maxar satellites. Pakistan encourages China to pursue similar claims for the destruction of its satellite. Maxar also publicly urges the State Department to file a claim under the Space Liability Convention for compensation for the satellites and to make a complaint to the World

Trade Organization alleging that Chinese and Pakistani actions constitute a non-tariff barrier to the use of space.

Pakistan, following through on its earlier notice to China, takes General Khan, CEO of the private entity using China's satellite services, in for questioning as part of the Pakistani investigation into the misuse of the Chinese-Pakistani satellite service. The United States praises this decision after it is leaked to the Pakistani press. At the same time, Pakistani leaders leak that its national security council has met and its military is preparing for readiness drills.

At this point, Maxar and India lose contact with their third satellite, including its U.S. payload, knocking out the constellation entirely. The Chinese government says that four-party talks between the United States China, India, and Pakistan are the only way to clarify what has occurred and head off the crisis, and Pakistan agrees. Shortly after, an Australian company releases imagery showing China's co-orbital rendezvous satellite actively engaged in a rendezvous-proximity operation with the third U.S.-India Maxar satellite shortly before it went offline. Pakistan, with whom China was not coordinating, learns of the attack for the first time from these public reports. Concerned about how continued escalation in space will affect its security, Pakistan urges China to refrain from further military action in space.

At the same time, the U.S. Cyber Command engages in offensive cyber attacks against China's co-orbital systems and anti-satellite laser infrastructure, intending to achieve hard kills against the assets China used to attack the U.S.-India Maxar assets. U.S. Cyber Command succeeds in disabling these assets. India and Pakistan are unaware of this covert action.

The United States, now satisfied that it has shut down China's ability to continue its anti-space actions in the near term, agrees to requests for four-party talks but requests a neutral location. Pakistan, initially unaware of the American cyber-attack, agrees and suggests a summit at the Secretary of State/Foreign Minister level in Dubai.

Given the rapid escalation of the crisis, Pakistan suggests holding the summit within five days to discuss building transparency in space activities, preventing the weaponization of space in the region, and creating agreed space-related crisis communication mechanisms, as well as the root issue of Kashmir.

China, however, rejects calls for talks because of the American disabling of the Chinese satellite. Beijing demands an American apology and renunciation of further use of force in space. This makes India and Pakistan aware of the attack for the first time. China says it will respond to U.S. hard kills on their assets through continual, proportional attacks on U.S. commercial satellites until the United States ceases its attacks. China's demand that Pakistan pre-clear its communications with Beijing is released via Wikileaks and picked up in the American press as evidence of Chinese hypocrisy.

India, meanwhile, remains focused on responding to the terrorist attacks against its nuclear forces and informs the United States of its intent to launch a cruise missile strike against the terrorist sites where the attack originated, which are in Pakistani territory. China intercepts and publicly releases India's plans and demands that India stand down, otherwise four-party talks will be impossible. India, however, had already launched the strike into Pakistani territory, which destroyed the terrorist site without civilian casualties.

The Indian Prime Minister then announces that India will also not engage in four-way diplomatic talks after China knocked out India's entire Maxar constellation. New Delhi accuses Beijing of creating debris that could jeopardize the International Space Station and other satellites owned by many different nations. India calls for a meeting of the UN Security Council. Meanwhile, the U.S. Space Command issues an enhanced defense posture to all U.S. government and commercial satellites and postures to protect vital U.S. defended assets. The United States reduces its broadcast of space domain awareness data which could be used for targeting and prepares to negate Chinese satellites, control centers, sensors, radars, and telescopes.

The Pakistan team began considering its options for a counter-strike against Indian targets proportional to India's missile strike against its territory but game play ended before it made any final decisions.

About the Authors

Brian Chow is an independent policy analyst and author of over 170 publications. He is a former Senior Physical Scientist at the RAND Corporation where he worked for 25 years. He was appointed by the President's Science Advisor, the Under Secretary of Defense for Policy, and the Office of the Chief of Naval Operations as a consultant to advise them on policy matters. His recent articles on space appear in *Strategic Studies Quarterly*, Center for Strategic and International Studies, *SpaceNews*, *The National Interest*, *Defense One*, *Defense News*, and *The Space Review*. He holds a Ph.D. in physics from Case Western Reserve University and an MBA (with Distinction) and Ph.D. in finance from the University of Michigan. He can be reached at brianchow.sp@gmail.com and on Twitter at @briangchow.

Peter Garretson is a Senior Fellow in Defense Studies at the American Foreign Policy Council and a strategy consultant who focuses on space and defense. He is the coauthor of *Scramble for the Skies The Great Power Competition to Control the Resources of Outer Space* and the host of AFPC's Space Strategy Podcast. Prior to joining AFPC, Col. Garretson spent over a decade as a transformational strategist for the Department of the Air Force, where he served as a strategy and policy advisor for the Chief of Staff of the Air Force, as Division Chief of Irregular Warfare Strategy Plans and Policy, and as the Chief of the Future Technology Branch of Air Force Strategic Planning. As an instructor of Joint Warfare at Air University, he laid critical foundations for the future of American and allied space-power, initiating the Schriever Scholars (America's premier program to develop space strategists), the Space Horizons Task Force (America's think tank for space), and developing the rationale for a U.S. Space Force. He taught courses in war theory, joint planning, and national security implications of artificial intelligence. Col. Garretson was the first serving U.S. officer to serve as a visiting fellow at India's premier strategic think tank, the Institute for Defense Studies and Analysis (IDSA) as a Council on Foreign Relations (CFR) international affairs fellow where he authored a vision for U.S.-India Space Cooperation. He was the designer of several Joint and interagency, airpower, and space-power wargames, including the Air Force Future Wargame Series (2005-2009) and the first-ever U.S.-UK-France Trilateral strategic wargame and first multi-agency asteroid deflection and disaster exercise. Garretson served as a collaborator in on numerous strategic documents, including the 2011 National Military Strategy, the first Air Force Unmanned Aerial Systems (UAS) Flight Plan, the Directed Energy Task Force, the Air Force Vision for Learning, Air Force Energy Horizons, Multi-Domain Command & Control Vision, DARPA 100 Year Starship, the Pentagon Study on Space Solar Power, the Fast Space Report for ultra-low-cost access to space, Air Force Space Command's Space Future 2060, and the Air Force's Strategic Environmental Assessment, Air Force Futures Group and Blue Horizons Program. A graduate of the U.S. Air Force Academy, Garretson received his Master's Degree in Human Factors in Aviation Science from Embry-Riddle University, and is presently pursuing a Doctorate in Public Policy at Auburn University.

Namrata Goswami is an independent scholar on space policy and Great Power Politics. Recently, she has been invited to teach at the Thunderbird School of Global Management, Arizona State University for their Executive Masters in Global Management, Space course. She is a consultant for Space Fund Intelligence and a guest lecturer at the seminar on "India Today: Economics, Politics, Innovation & Sustainability," Emory University. She was subject matter expert in international affairs with Futures Laboratory, Alabama. She worked as Research Fellow at MP-Institute for

Defence Studies and Analyses, New Delhi; a visiting Fellow at Peace Research Institute, Oslo, Norway; La Trobe University, Melbourne, Australia; University of Heidelberg, Germany; Jennings-Randolph Senior Fellow, United States Institute of Peace; and was a Fulbright Senior Fellowship Awardee. She was awarded the Minerva grant by Office of the U.S. Secretary of Defense to study great power competition in outer space. In April 2019, Dr. Goswami testified before the U.S-China Economic and Security Review Commission on China's space program. Her co-authored book, *Scramble for the Skies: The Great Power Competition to Control the Resources of Outer Space* was published 2020 by Lexington Press; Rowman and Littlefield. Her book on *The Naga Ethnic Movement for a Separate Homeland* was published in 2020 by Oxford University Press. She has published widely to include in *The Diplomat, the Economic Times, The Washington Post, Ad Astra, Asia Policy, Live Encounters Magazine,* and *Cairo Review*. She was invited in November 2019 to share about her life and her work at a Tedx event held at the Rosa Parks Museum, Montgomery, Alabama. She is currently working on two academic book projects, one on *China's Grand Strategy and Notions of Territoriality* (Routledge, 2023) and the other on *Spacepower Theory and Practice: Case Studies of U.S. China, India, Russia and Japan* (2023).

Peter Hays is a retired Air Force Lt Col who works as a defense contractor in the Pentagon supporting the Policy, Strategy, and Plans Directorate (S5S) under the Chief Strategy and Resourcing Officer on the Space Staff. He has been directly involved in developing and implementing most of the major national security space policy and strategy initiatives since 2004. Professor Hays' research interests coincide with his current graduate seminars on "Space and National Security" and "Science, Technology, and National Security Policy" at the Space Policy Institute at George Washington University, served as the Space Chair at Marine Corps University (MCU), and teaching air-and space-power seminars at the MCU School of

Advanced Warfighting. He was a 1979 Honor Graduate of the USAF Academy and earned a Ph.D. from the Fletcher School. Hays' major publications include *Handbook of Space Security, Space and Security,* and *Toward a Theory of Spacepower*.

Gregory Jones has served as a defense policy analyst for the past forty-nine years. Over the course of his career, a major emphasis of his work has been the study of the potential for terrorists as well as countries to acquire and use nuclear, chemical, biological and radiological weapons, and the formulation of policies and actions to control and counter these weapons. He has been a senior research analyst at Pan Heuristics, RAND, and the Nonproliferation Policy Education Center. Mr. Jones is currently an independent consultant and publishes the website Proliferation Matters. He is the author of the book *Reactor-Grade Plutonium and Nuclear Weapons: Exploding the Myths* (Nonproliferation Policy Education Center, 2018), a coauthor of the book *Swords from Plowshares* (The University of Chicago Press, 1979), and the author or coauthor of over one hundred reports and articles regarding defense policy.

Brandon Kelley is the Associate Director of Debate at Georgetown University, and is currently pursuing a M.A. degree in Military Operations from Georgetown's Security Studies Program (Spr. '22). He began conducting open source research and intelligence work regarding Chinese and Russian counterspace capabilities while an undergraduate. Following graduation in 2017, he was contracted to do much of the research and writing for the inaugural edition Global Counterspace Capabilities: An Open Source Assessment, now one of the primary resources for information on global anti-satellite capabilities. In 2020, he co-authored a report, published by the Nonproliferation Policy Education Center, regarding the legality and desirability of "safety zones" as a solution to co-orbital ASAT threats in GEO. His primary research and career interests center around strategic deterrence dynamics, military strategy and operations, and finding solutions to next-generation ASAT threats. Brandon currently coaches and administers the multiple-time national championship-winning Georgetown Policy Debate Team full-time. He holds a B.S. in Science, Technology, & Int'l Affairs from Georgetown University's School of Foreign Service.

Laura Montgomery is an adjunct professor of space law at Catholic University's Columbus School of Law, and the proprietor of Ground Based Space Matters, LLC, where she specializes in regulatory space law, with an emphasis on commercial space transportation and the Outer Space Treaties. Ms. Montgomery spent over two decades with the Federal Aviation Administration. During her tenure there, she served as the manager of the Space Law Branch in the FAA's Office of the Chief Counsel.

Henry Sokolski is executive director of the Nonproliferation Policy Education Center and teaches graduate classes on nuclear policy at the University of Utah and the Institute of World Politics. He was recently appointed Senior Fellow for Nuclear Security Studies at the University of California at San Diego. Previously, he worked in the Pentagon as Deputy for Nonproliferation Policy, as a consultant to the National Intelligence Council, as a member of the Central Intelligence Agency's Senior Advisory Group, as a Senate military legislative aide to a member of the Senate Armed Services Committee, and as a special assistant on nuclear energy to the chairman of Senate's Tennessee Valley Authority oversight committee. He served on two congressional commissions on the prevention of WMD proliferation and has authored and edited numerous volumes on strategic weapons proliferation issues, including *Best of Intentions: America's Campaign against Strategic Weapons Proliferation* and *Underestimated: Our Not So Peaceful Nuclear Future.* Mr. Sokolski earned his graduate degree in political science from the University of Chicago.

Sam Wilson is a senior policy analyst for the Center for Space Policy and Strategy at The Aerospace Corporation. In this role, he is responsible for leading work on the nexus of commercial and national security space, comparative space, and missile issues. Wilson has authored papers covering the technical characteristics of the broader missile environment, the convergence of commercial and national security space, the everyday uses of space systems, the impact of Brexit on the U.K. space sector, Japan's shift toward a focus on space security, the U.S. hypersonic missile debate, the importance of international space partnerships, and space traffic management. His work has appeared or been covered in *The Washington Post, Politico, SpaceNews, Breaking Defense, Kyodo News, Japan Today, The Bulletin of the Atomic Scientists, The Diplomat, Milsat Magazine,* and *Satellite Magazine*, among other outlets. Prior to joining Aerospace, Wilson served as a senior defense analyst at the U.S. Government Accountability Office. There, he led reports on strategic force structure; arms control; nuclear command, control, and communications; and U.S. nuclear forces in Europe. He also drafted statutory language, including sections in the National Defense Authorization Act, at the request of congressional staff. Wilson has completed fellowships with the National Defense University's Program for Emerging Leaders, the Center for Strategic and International Studies Project on Nuclear Issues, the German Marshall Fund, and the Nonproliferation Policy Education Center. Wilson received his bachelor's degree in political theory and master's degree in public policy from the University of Virginia, where he graduated with distinction.

Brent Ziarnick, PhD, is an associate professor in the Department of Spacepower at the Air Command and Staff College, Maxwell AFB, Alabama. There he teaches military theory and space-power strategy to US Space Force and sister service midcareer officers. He developed and founded the Schriever Space Scholars, the only space-centric, year-long professional military education program in the world. He is also an Air Force Reserve space operations officer, having served as a squadron commander, deployed space control officer, and Air Force Space Command strategist. Dr. Ziarnick has authored three books and over a dozen articles on space topics and is a frequent contributor *to The Hill, Politico*, and *Air Force Times.* He holds a master's degree in space systems engineering and doctorates in economic development and military strategy

Made in the USA
Monee, IL
03 November 2026

67374648R00203